EARTH
CANCER

EARTH
C A N C E R

Van B. Weigel

PRAEGER

Westport, Connecticut
London

Library of Congress Cataloging-in-Publication Data

Weigel, Van B.
    Earth cancer / Van B. Weigel.
      p. cm.
    Includes bibliographical references and index.
    ISBN 0–275–95177–4 (alk. paper)
      1. Environmental ethics.   2. Man—Influence on nature.   3. Human
ecology. I. Title.
GE42.W45   1995
304.2'8—dc20   95–2211

British Library Cataloguing in Publication Data is available.

Library of Congress Catalog Card Number: 95–2211
ISBN: 0–275–95177–4

First published in 1995

Praeger Publishers, 88 Post Road West, Westport, CT 06881
An imprint of Greenwood Publishing Group, Inc.

Printed in the United States of America

The paper used in this book complies with the
Permanent Paper Standard issued by the National
Information Standards Organization (Z39.48–1984).

10 9 8 7 6 5 4 3 2 1

Copyright Acknowledgments

The author and publisher gratefully acknowledge permission for the use of the following material:

Reprinted by permission of the publishers *Origins of the Modern Mind: Three Stages in the
Evolution of Culture and Cognition* by Merlin Donald, Cambridge, Mass.: Harvard University Press,
Copyright © 1991 by the President and Fellows of Harvard College.

Reprinted by permission of the publishers *Biophilia* by Edward O. Wilson, Cambridge, Mass.:
Harvard University Press, Copyright © 1984 by the President and Fellows of Harvard College.

*For Linda . . . whose passion for life
is exceeded only by her
intelligent love*

Contents

Preface

Earth, an exceptional and remarkable planet, an island of life within a barren universe, has cancer. The truth of this is too painful to grasp.

The *Homo sapiens*—a proud, intelligent life form and recent arrival to the planet—is out of control. Like normal cells within a healthy organism that become cancerous and devour life around them, human beings have unwittingly trapped themselves into life modes that are destroying the virtual organism of life that we call the biosphere. The spread of earth cancer is seen not through x-rays but through satellite imagery. Where healthy tissues remain, humans lie in wait.

It is not that human beings are cancerous by nature. Just as the potential for cancer lies within the normal processes of healthy cells, earth cancer is the result of a malignant imbalance between the aspirations of human beings to colonize the planet and the aspirations of other life forms to thrive.

Recent studies on cancer in humans and animals emphasize how the seeds of cancer are present within the normal processes of healthy cells. When the normal regulatory mechanisms of an organism go awry, perfectly healthy cells have the potential to become cancerous. For example, the development of malignant tumors appears to be related to the failure of cells to die when their time comes. A cell that outlives its programmed death, because of miscued protein signals from its genetic code, adds to the mass of the tumor (Marx, 1993). A similar dynamic seems to be related to the development of secondary tumors (or metastases) within an organism—the most life-threatening aspect of cancer. For cancer to spread within an organism, malignant cells must have the capacity to migrate and colonize distant organs by invading healthy tissues.

This invasive capability is not only a property of cancer cells but also a capacity of normal cells. Normally the regulatory mechanisms of the body restrain such cellular invasions. When the environmental stimulus that initially provoked the invasion is removed, the cell halts. Not so with unrestrained malignant cells. Restless and indiscriminate marauders, they penetrate tissue barriers in contexts that would be improper for normal cells (Liotta, Steeg, and Stetler-Stevenson, 1991; Liotta, 1992).

An invasive and unrestrained malignant cell is to human cancer what an invasive and unrestrained human being is to earth cancer. Both represent an aberrant version of their original constitution; both are permitted to thrive because of a systemic imbalance; both are deadly to the health of the organism.

Humans are curious animals. On the one hand, they claim for themselves tremendous intelligence and insight. On the other hand, they have difficulty accepting a truth as basic as the interdependent nature of their existence. Indeed most modern manifestations of human thought and aspiration have persistently and artfully denied the fact of this interdependence, malignantly placing ultimate value on the worth of human prerogatives and achievements.

This book is a challenge to the absolute priority accorded to human goals and aspirations. Its thesis is simple. If humans are truly an intelligent species, they must define their economic, social, and moral values in a manner that is benign—consistent with the essential interdependence of the planet we call Earth. Stated alternately, earth cancer must be stopped dead in its tracks. The malignancy must end. Action follows self-understanding. Humans have a profound responsibility to assume their proper role as biospheric agents of remission.

The recognition of the essential interdependence of the biosphere appears to be both a self-evident *and* unremarkable truth. Yet, as I endeavor to show throughout the book, the interdependent nature of our existence may be self-evident, but it is hardly unremarkable. On the contrary, the recognition of our connectedness with the earth's ecology cannot help but have a profound effect on how humans think of themselves and define their place in the world. The old thinking of imperialistic superiority and subjugation must give way to the new thinking of respect for all life forms and an ethic of interdependence.

An ethic of interdependence, if it is to reflect the dynamic character of ecological interdependence, can neither avoid the harsh realities of predation nor adopt an unqualified sacred status for all life forms. Death and birth, competition and cooperation are defining features of life within Earth's biosphere. Interdependence does not preclude pain and suffering; it cannot be "managed" in the usual sense of the term. To affirm the worth of ecological interdependence is to embrace the struggle of life around us. The "balance of nature" is not a fixed, fragile state of harmony but the dynamic process of conflict and change.

The destiny of the human species is inextricably linked to an intricate web of biological relationships which we call the biosphere. Our relationship with other life forms is so close that the respective fates of hundreds of thousands of species are directly contingent upon human decisions. In modern times the impact of human predation upon other animal species is minor in comparison to the nonpredatory actions of humans—manifest most directly in the destruction of habitats and the prolific generation of waste. Instead of finding our place within the biosphere, we have behaved like shiftless pirates roaming the planet in search of ecological booty.

The premise of this book is that humans have an ecological imperative to recover our heritage as "members in good standing" of Earth's biosphere. To do so, humans must discover anew the meaning of ecological interdependence and forswear their participation in earth cancer. For ease of reference, I refer to this ecological mandate as the Prime Directive: Respect the essential interdependence of the biosphere. Unlike the Prime Directive of Star Trek fame, this is *not* a principle of noninterference. *Homo sapiens*, as long as they exist, will always impact the biosphere. We can no more deny our influence upon the web of life than any other creature. Instead the Prime Directive referred to here requires humans to live in a way that is consistent with the interdependence of the biosphere, instead of diminishing or destroying it.

It would be difficult to overstate the importance of the Prime Directive for modern life. We live in a time of unprecedented environmental devastation and species extinction—all due to the decisions of one species, the *Homo sapiens*. Each and every day approximately 75 plant and animal species are driven into biological extinction because of human decisions—a rate of extinction that is at perhaps 10,000 times the recent historical rate of extinction (Wilson, 1992). What intelligent life form would consider their goals and aspirations to be so noble and exalted that they could justify such widespread devastation? Yet, we not only persist in wreaking havoc and destruction upon our planet's ecology, but we also have dreamed up exotic rationalizations to legitimate our behavior.

Humans have long fancied themselves to be at the center of *the* universe. In fact, we are only at the center of *our* universe. Many humans believe that God created the planet, if not the universe itself, as a kind of private preserve for human enjoyment and development. Similarly, those who pride themselves as children of the Enlightenment, casting off the chains of religious superstition, have indulged in anthropocentric (human-centered) fantasies which rival or exceed those of their religiously inspired predecessors.

Modernization, technological progress, and economic development are convenient terms for the transformation and subjugation of the earth's ecology for human purposes. With incredible arrogance we define the planet's natural endowment of water, air, soil, minerals, fossil fuels, and the like as "natural resources" or "factors of production," as if they exist expressly for our use and exploitation. Even those who champion ecological causes typically succumb to the human-centered calculus that environmental protection is important because it is important for human well-being—for present as well as future generations. The idea that the Earth's biosphere has autonomous value—independent of human preferences—is fundamentally foreign to a way of thinking that exalts exorbitant expansion, mindless migration, and consumptive colonization. The courage, compassion, and insight of humans is not dignified by such things. It is not our destiny to live as cancer cells. This book is written toward that end.

Acknowledgments

I would like to extend my sincere appreciation to the following individuals for their support on this project: Lyndell Backues, Peter Genco, David Gibson, Christina Jackson, Julia Link, Thomas McDaniel, Anna Montaney, Elizabeth Morgan, John and Miriam Nolte, Grant Power, Joseph Sheldon, Phillip Thomas, Norma Thomasson, John Whealin, David Wilcox, and my parents, Robert and Donna Weigel.

Chapter 1
Living Without Berlin Walls

THE WALL MUST COME DOWN

"I think, therefore I am." These famous words of René Descartes have immortalized the Berlin Wall between the creatures called *Homo sapiens* and all other life forms on planet Earth. For Descartes, animals were machinelike "thoughtless brutes" who possessed no conscious life whatsoever—not even an awareness of pleasure and pain. Only humans, those creatures capable of ratiocination, could genuinely experience life.

Like the real Berlin Wall, the Cartesian wall between humans and all other creatures is destined for the ash heap of history. *Cogito, ergo sum* is slowly giving way to *vivo, ergo sum* ("I live, therefore I am"). Human life must necessarily define itself within a planetary web of irrepressible interdependence. Uniqueness can no longer be understood as separateness; the capacity for rational thought can no longer be a justification for the cancerous colonization of life.

Earth cancer could not exist without Berlin Walls. Ecological totalitarianism, not unlike political totalitarianism, relies on ignorance and isolation, prejudice and powerlessness to further its aims of crippling control and colonization.[1] Just as political totalitarianism has been possible only within the twentieth century, with the elaborate means of surveillance and domination that modern technology affords, so also ecological totalitarianism has been perfected into a technological art form during this century.

This book is about the crumbling of Berlin Walls and the demise of ecological totalitarianism. Berlin Walls are ideally suited to reproduce the requisites of totalitarianism, but Berlin Walls do not last forever. Once the subversive literacy of empowerment gains a foothold, it is only a matter of time before Berlin Walls lose their legitimacy and are dismantled by people of courage who have eyes to see.

Those who climbed the real Berlin Wall on November 10, 1989—tooting trumpets and dancing atop the wall—experienced an unparalleled sense of joy and exhilaration. West Berlin, by 3:00 A.M., assumed the atmosphere of a giant block party. These heralds of a new era brought not only champagne but also hammers and chisels. One young woman from East Berlin, with tears in her eyes, exclaimed "I can't describe it; I would never have believed it possible." A middle-aged East Berliner captured his feelings with the words, "Joy, entirely great joy" (Protzman, 1989).

An even more profound sense of elation awaits those who triumphantly tear down the ecological Berlin Wall that obstructs the "reunification" of the biosphere. While anxiety and concerns about the costs of ecological reunification may dampen momentarily the enthusiasm sparked by crumbling walls, as with the reunification of Germany, there can be little doubt that humanity, like Germany, is better off without Berlin Walls. Crumbling walls give way to new horizons, new opportunities. Landscapes once characterized by dull monotony are transformed into colorful variety; bland predictability gives way to the wonder and mystery of life; things once divided are restored to a robust symmetry.

Humans, as a species, are premier wall builders. One of the enduring architectural legacies of the human occupation of planet Earth is the Great Wall of China—the only human-made structure that is visible with the naked eye from the moon. It is a monument to our social and political need for boundaries.

Tragically, the walls that we have erected on the landscapes of our planet have also been etched into our minds. The collective impact of culture and the history of ideas in the West has erected a Berlin Wall that has impoverished our species to the extent that it threatens our very existence. It is a wall that has nourished a death wish within the human heart. It is a wall that has encouraged our species to become heartless angels of extinction, triumphantly reenacting the legacy of Adolph Hitler and Pol Pot within the ecological sphere. It is a wall that must come down.

BERLIN WALLS AND DEVASTATING DEVELOPMENT

Human communities of the twentieth century have been abducted by cancerous colonization. Progress has become a euphemism for a virulent and self-maiming form of narcissism. Economic development has become largely synonymous with increased consumption, industrialization, and urbanization. Environmental devastation has come to be accepted as mere "collateral damage" on the pathway to progress.

Devastating development is the logical outcome of the transformation of Economic Man (*Homo oeconomicus*) into Cancerous Man (*Homo canceris*). Economic development, for instance, is all too frequently associated with urbanization, the clearing of virgin forests, reliance on mechanized forms of transport, and the overexploitation of nonrenewable resources. While development theorists frequently talk about "sustainable development," this term functions more as a rhetorical catchword than a meaningful concept that embodies clear policy directives (Lele, 1991). Economic theory and development policy are still captive to the mantras of modernization theory—an unimaginative approach to development that amounts to little more than a game of "follow the leader" among nation-states. Modernization is mimicry. Pollution is a proxy for progress.

The concept of modernization, like that of progress, is fraught with incapacitating contradictions. The less developed countries of the world are less developed because they have not attained the standard of living of the rich countries—a standard of life that sets humanity on a perilous pathway toward an ecological Hades. As a result, development has become virtually synonymous with the production of waste materials on a vast scale, crippling overpopulation, ecological devastation, obscene social inequalities, urban congestion, and perpetual pollution. This is "development" at the end of the twentieth century—devastating development.

DEVELOPMENT AND THE EXTINCTION OF LIFE

Under the benevolent, watchful eye of human progress, the Berlin Wall between humans and all other life forms has masked an unprecedented era of extinction across the planet. One compelling evidence of earth cancer is that approximately 27,000 plant, insect, and animal species are doomed to extinction each year within rain forests *alone*—a conservative estimate. This amounts to the extinction of 74 species each day. The current rate of extinction may be as much as 10,000 times the natural (or background) extinction rate for the planet (Wilson, 1992: 280). It is likely that it will take somewhere between 10 and 25 million years (or between 400,000 and 1,000,000 human generations) before the natural evolutionary processes of speciation can rectify our generation's devastation of Earth's biodiversity (Myers, 1991). Table 1.1 is a telling portrait of a trail of destruction left by human communities of our day.

As Richard Leakey and Roger Lewin point out, the current tidal wave of extinction is unprecedented not because of its *magnitude* but because of its *cause* (1992: 348, 354–55). Of the five mass extinctions in the history of our planet, the Permian extinction, occurring some 225 million years ago, was clearly the most devastating. Ninety-six percent of all Earth's species were destroyed! More recently, the Cretaceous extinction, occurring 65 million years ago, marked the end of the majestic rule of dinosaurs and the extinction of from 60 to 80 percent of all terrestrial species. For over 150 million years, the dinosaurs dominated the landscapes of our planet—nearly 40 times longer than the comparatively brief interval between us and the beginnings of the hominid family, some 4 million years ago.

Yet, unlike the Cretaceous extinction and those before it, the cause of today's mass extinction is not a wayward asteroid or some other temporary natural disturbance. We are the cause. The miraculous capacity of the Earth to regenerate itself will be severely impeded by our continued cancerous presence on the planet. Self-annihilation is obviously neither a desirable nor a feasible solution to the problem. The solution instead lies in rediscovering our humanity—affirming both the value of human life and the value of all life.

THE PATTERN OF PROGRESS

Human ''progress'' is neither progress for humanity nor progress for the planet. It is a curiously twisted mixture of naive, arrogant, and self-destructive assumptions that legitimate our role as Earth's cancer agents.

Human progress in 1990 alone destroyed tropical forests equal in size to the state of Washington (Worldwatch Institute, 1992). In effect, this earmarked 1.3 acres of tropical forest for destruction each second! Not only does this wholesale devastation devour the irreplaceable DNA of thousands of plant and animal species, but also it intensifies the growing problem of carbon dioxide buildup within our atmosphere. While we

Table 1.1
Observed Declines in Selected Animal Species, Early 1990s

Species Type	Observation
Mammals	Almost half of Australia's surviving mammals are threatened with extinction. France, western Germany, the Netherlands, and Portugal all report more than 40 percent of their mammals as threatened.
• Primates	More than two thirds of the world's 150 species are threatened with extinction.
• Carnivores	Virtually all species of wild cats and most bears are declining seriously in numbers.
Birds	Three fourths of the world's bird species are declining in population or are threatened with extinction.
Amphibians	Worldwide decline observed in recent years. Wetland drainage and invading species have extinguished nearly half of New Zealand's unique frog fauna. Biologists cite European demand for frogs' legs as a cause of the rapid nationwide decline of India's two most common bullfrogs.
Reptiles	Of the world's 270 turtle species, 42 percent are rare or threatened with extinction.
Fish	One third of North America's freshwater fish stocks are rare, threatened, or endangered; one third of U.S. coastal fish have declined in population since 1975. Introduction of the Nile perch has helped drive half the 400 species of Lake Victoria, Africa's largest lake, to or near extinction.
Invertebrates	On the order of 100 species are lost to deforestation each day. Western Germany reports one fourth of its 40,000 known invertebrates to be threatened. Roughly half of the freshwater snails of the southeastern United States are extinct or nearly so.

Source: Adapted from Worldwatch Institute, *State of the World, 1992* (New York: W. W. Norton, 1992), p. 13.

cannot be sure about the long-term impact of CO_2 pollution on the Earth's climate, the prospect of global warming cannot be handily dismissed as just one more scare tactic of the environmental movement. Of particular concern is the uncertain impact of global warming on ocean currents which moderate extremes in the Earth's climate. If, for example, greenhouse warming prevents the formation of North Atlantic "deep water," thereby stalling the northward flow of the warm Gulf Stream, the climatic impact would be severe and swift—making it impossible for many species to adapt to the new conditions (World Resources Institute, 1994: 206–7).

Human progress, because of its narrow focus, cannot easily grasp the profound connection of all life on the planet. Take, for example, the impact of chlorofluorocarbons (CFCs) upon the ozone layer. We can hardly fathom the consequences of a depleted ozone layer for Earth's life forms. The increased incidence of skin cancer and eye damage among humans would be among the more benign outcomes. More perversely, ozone depletion has the potential to mutate the Earth's oceans into vast "dead zones." Increased levels of UV-B radiation can reduce the productivity of or even kill small organisms that dwell on the ocean's surface (e.g., phytoplankton, zooplankton, fish larvae), interrupting the complex food chain that governs life under the sea (Hardy and Gucinski, 1989). Recent studies conducted in Antarctica during the peak of the ozone hole show that increased UV radiation has caused a seasonal decline in the production of vegetative plankton of from 6 to 12 percent (Schneider, 1991; World Bank, 1992: 63).

What definition of progress can sanction such indiscriminate and widespread devastation in the furtherance of human civilization? Our effort to remake the biosphere according to the dictates of comfort and convenience has the effect of making it inviable for other life forms—having the cumulative effect of making it inviable for us as well.

A TALE OF THREE CITIES

One need only visit the world's major urban centers to witness the impact of devastating development. Mexico City, the most populous urban center on the planet, has become a smog-saturated crater, choking its population with air pollution that numbs not only the senses but also the mind. A study in 1988 indicated that the incidence of lead poisoning among more than half the newborn babies in Mexico City was substantial enough to impair their neurological and motor-physical development (Rothenburg et al., 1989). Among adults it is estimated that nearly 20 percent of the incidence of hypertension in Mexico City—a contributing factor to heart attacks and strokes—is caused by increased exposure to lead (World Bank, 1992: 53). Rosario Camacho, a 44-year-old Mexican housewife, does not need to follow the government's air quality index to understand the severity of the problem. She can tell the onset of the annual pollution season by the fact that her four children cannot sleep at night because of respiratory problems (Cody, 1991).

The mind-numbing effects of lead exposure are also felt in other population centers, such as Bangkok, Thailand. It has been estimated that children in Bangkok experience an IQ loss of four or more points by age seven due to increased exposure to lead (World Bank, 1992: 53). While U.S. health authorities consider 5 micrograms of lead per deciliter of blood to be hazardous for children (and 25 micrograms dangerous for adults), hospitals in Bangkok characterize a level of 35 micrograms–the average for long-term residents—as "acceptable." Moreover, the incidence of lung cancer is three times higher in Bangkok than in Thailand's outlying provinces (Magistad, 1991). Indeed, Bangkok, Beijing, Calcutta, New Delhi, and Tehran have exceeded the air pollution limits established by the World Health Organization for more than 200 days a year—limits that should not be exceeded for more than 7 days a year (World Bank, 1992: 50).

Manila, once known as the "pearl of the Orient," has become a kind of urban paradigm for devastating development. Poverty, overpopulation, unrestrained industry, and inadequate planning have conspired to create a city of stench. The once graceful Pasig River, which runs through the city, has become a communal latrine for industrial waste, garbage, and raw sewage. A ferry trip down the Pasig is an adventure in septic ooze: The waters are swollen with garbage, milky fluids spew from open pipes, and the stench of raw sewage is staggering. Manila's five other rivers are also biological wastelands. One official from the U.S. Environmental Protection Agency surveyed Manila's Tullahan River and "got sick and puked all day." The quality of Manila's air—approaching conditions like those in Mexico City—is no better than that of its rivers and canals. The problem is that no one knows precisely how bad things are. The air-monitoring equipment for the city, donated in 1974, broke down and was never replaced. Manila's air quality was last measured in 1983. There is, however, one very reliable measure of the city's air pollution:

Asthma and other respiratory diseases are rampant among Manila's children (Broad and Cavanagh, 1993: 20ff; Drogin, 1990).

THE SMOKESTACK AND THE SICKLE

Significantly, the pattern of devastating development applies equally to both laissez faire capitalist and command socialist modes of development. Nearly 80 percent of all manufacturing in the former Soviet Union took place within the Russian Republic—a fact that may explain why the life expectancy in Russia is shorter than in any other European country, including the Ukraine, Byelorussia, and the Baltic States. Sixty-eight of the industrial centers of the former Soviet Union have levels of air pollution that are hundreds of times higher than the permissible levels established by the government (*The Economist*, 1989).

If Manila is a paradigm for capitalist-inspired devastating development, Magnitogorsk is its communist counterpart. A Stalinist creation of the 1930s in the southern Urals, Magnitogorsk is home to the Lenin Steel Works, the largest steel mill in the world. Unfortunately for its 430,000 inhabitants, Magnitogorsk is also home to some of the most venomous smokestacks on the face of the Earth. This "Steeltown U.S.S.R." belches a million tons of filth into the air every year—more than two tons for every man, woman, and child in Magnitogorsk. Snow turns black in the winter; otherwise green grass takes on brownish hues in the summer. Ninety percent of Magnitogorsk's children eventually suffer from some type of pollution-related illness: chronic bronchitis, asthma, allergies, and cancers. The situation is so bad that the government has set up a "prophylactic clinic" where children can receive ultraviolet-light treatments and "oxygen cocktails" to make it through the day (Remnick, 1991).

Altogether more than 20 million Russians are forced to breathe and drink seriously contaminated air and water. Legally mandated air and water pollution standards are routinely disregarded. Approximately three-fourths of Russia's surface water is classified as not fit to drink. A study in 1989 found that 69 percent of the freshwater fish were extremely contaminated by mercury-based pesticides. Numerous beaches along the Azov, Black, and Baltic seas have been routinely closed for swimming in recent years (Yablokov et al., 1991).

The Aral Sea is a case study in biological devastation, Soviet-style. It was once the fourth largest lake in the world. During the brief space of twenty years, thanks to the diversion of its feeder rivers for cotton production in Uzbekistan, it has lost one-half of its water and is now biologically dead. The last fish died in 1983. Once thriving fishing villages have been long since abandoned and are now 40 miles from the shrinking sea. Every year the wind deposits 75,000 tons of salt and dust on the cotton fields of Uzbekistan from the parched seabed of the Aral—in poetic justice, poisoning the very fields for which the Aral was sacrificed (*The Economist*, 1989).

ERODING THE GOOD EARTH

Unfortunately the Aral Sea and the fields of Uzbekistan are but two casualties of devastating development. In 45 brief years, our species has managed to seriously degrade the soil over a landmass the size of China and India combined, by employing damaging agricultural practices which promote erosion and the loss of nutrients, industrialization, overgrazing, and deforestation (World Resources Institute, 1992: 111ff). If we were to compress the time between us and the invention of agriculture into a calendar year, assigning January 1 to the beginning of agriculture, we have managed to degrade 11 percent of the Earth's soil from 3:00 PM on December 30 to midnight on December 31.

The tiny Central American country of El Salvador, roughly the size of the state of Massachusetts, has suffered under the dual burden of protracted war and ecological degradation. Once a lush and fertile land, El Salvador is now an ecological nightmare. Ninety-four percent of the country's original forest has been destroyed, along with 80 percent of its natural vegetation. Altogether, 77 percent of El Salvador's soil has been either seriously damaged or made useless by the ensuing erosion. The productive land that remains is laden with dangerous pesticides such as DDT. These pesticides, combined with other toxic chemicals, have found their way into the food chain, threatening not only many species of El Salvador's flora and fauna but also its human population. A once-fertile marshland covering 290,000 acres has lost 75 percent of its plant life. An average of 50 children die each year as a result of pesticide poisoning in Children's Hospital of San Salvador alone (Freed, 1991). These problems, of course, are by no means confined to El Salvador. In nearby Nicaragua and Guatemala researchers have found the highest concentrations of DDT ever recorded in human breast milk (World Bank, 1992: 140).

An especially poignant portrait of environmental degradation comes from Madagascar, an island-nation the size of the state of Texas located

300 miles off the East African coast. The island has the distinction of
being one of the world's twelve "megadiversity" countries which collectively make up 70 percent of our planet's biodiversity. Madagascar is
home to the most unique primate population in the world. Twenty-eight
of its thirty primate species are unique to the island; four of the five
families of primates inhabiting Madagascar are found nowhere else in
the world. Furthermore, about 80 percent of Madagascar's plants, 88
percent of its reptiles, 98 percent of its amphibians, and 42 percent of
its bird species are unique to the island. In short, the island is a virtual
treasure trove of biological life. The island is home to the rosy periwinkle
(*Cartharanthus roseus*)—an inconspicuous plant that is the only known
cure for some forms of leukemia in children and has proven extremely
effective in the treatment of Hodgkin's disease in young adults. None of
the five other species of *Cartharanthus* that are unique to Madagascar
have been studied; one is in danger of extinction.

Madagascar is also home to a human population trapped by grinding
poverty. Most who cultivate the land do not own it and, consequently,
have little interest in preserving the soil. Moreover, the pressure of the
growing Malagasy population has fueled the destruction of the island's
rain forests by means of slash-and-burn agriculture. The island's ancient
forests are felled to make charcoal, hillsides are scorched to clear land
for a rice culture known as tavy, and flatlands are burned in order to
stimulate the growth of new grass for cattle to consume. Thus far humans
have been responsible for the extinction of a huge elephant bird (*Aepy-
ornis maximus*), two species of giant tortoises, a pygmy hippopotamus,
an aardvark species, and 14 species of lemur. Altogether, 85 percent of
Madagascar's forests have been cut down; 80 percent of the island's
energy needs are being met by burning wood.

At the present rate of population growth and deforestation, one estimate holds that Madagascar's forests will be lost forever in only 15 to
20 years (World Resources Institute, 1994: 247). Although Madagascar
is one of the few countries of the world with a National Environmental
Action Plan (implemented in 1991), the prospects for meaningful change
are bleak, as the island's environmental problems stem from the land use
decisions of a geographically dispersed population (Larson, 1994). The
current rate of deforestation not only has placed many unique species at
risk of extinction, but also has precipitated a massive loss of topsoil
which is being dumped into the Indian Ocean. One Soviet cosmonaut
who visited Madagascar commented to his hosts that he could always
identify the island from space because it was surrounded by a bright red

halo—an effect created as the topsoil washed out to the sea (Bartlett, 1991; Hiltzik, 1989; Masland, 1989; McNeely et al., 1990: 95; Wilson, 1991a).

The devastating development that creates bright red halos in the ocean, contaminates breast milk, blackens snow, renders seas lifeless, and lowers the IQs of children is cancer to us and our planet. Devastating development hardly befits the intelligence of our species. How could an intelligent life form like the *Homo sapiens* forfeit their common sense to become *Homo canceris*? Why would we entertain a self-destructive definition of progress that requires elaborate self-deception and demands the erection of a Berlin Wall between humans and all other creatures? Are we condemned to live in the shadow of the Berlin Wall, or is there another way?

To come to terms with these and other questions, we must first understand more about human beginnings. We cannot envision our future without first apprehending our past. We explore these questions in the next chapter.

NOTE

1. It is one of the ironies of the environmental movement that the term "eco-fascism" has been employed against "deep ecology" approaches to environmental responsibility. Those who invoke this label forget the legacy of ecological totalitarianism that typically characterizes their own position.

Chapter 2

Back to the Future

CHILLING OUT ON PLANET EARTH

In June 1992, an unprecedented gathering of world leaders converged on Rio de Janiero for the Earth Summit—a historic meeting that reflected increasing concerns about the problem of global warming due to CO_2 pollution and the precipitous decline in Earth's biodiversity. It is a curious irony of our planet's history that the very species that may cause a dramatic warming trend in the planet's climate, because of its prolific capacity for waste, is indebted to periods of global cooling for its own evolutionary development.

Speculation about human origins is admittedly a highly tentative enterprise. In this regard the humility embodied in the biblical phrase—"We see through a glass darkly"—is very a propos. With this qualification, it is likely that the emergence of the hominid family, some 4 million years ago, as well as the later emergence of the genus *Homo*, about 2.5 million years ago, coincided with successive periods of global cooling. As Richard Leakey surmises,

> What of the origin of the genus *Homo*? Does it "coincide" with anything significant? Yes, it does. . . . Huge ice mountains built up in Antarctica close to 2.6 million years ago, and for the first time significant amounts of ice formed in the Arctic. The frigid grip produced cooler, drier climates in the rest of the globe, including the varied highland terrain of eastern Africa. (Leakey and Lewin, 1992: 164)

These climatic changes broke up stable habitats, creating new opportunities for the development of new species. It is significant, for example,
that existing species of African antelopes suddenly vanished about 2.6
million years ago, to be replaced with a range of new antelope species.

Certainly the most striking feature of our evolutionary history is the
significant growth in the size of the human brain. In fact, the growth of
the hominid brain is a defining feature of the genus *Homo*. Extant fossil
evidence provides eloquent testimony of this remarkable development.
Whereas the mean value of the cranial capacity of extant Australiopi-
thecine skulls is only 464 cm^3 (compared to a range of from 300 to 480
cm^3 for chimpanzees), extant *Homo habilis* skulls have a mean size of
657 cm^3 and *Homo erectus* averages 978 cm^3. The mean brain size of
Homo sapiens is 1,300 cm^3 with a range of from 1,000 to 2,000 cm^3
(Crook, 1980: 128–29). Of particular importance was the visible expansion of the frontal lobes of the brain (which include the Broca's area, a
primary language center)—a development that required the formation of
the unusually high forehead that characterizes human craniums (Leakey
and Lewin, 1977: 192, 205). Significantly, the oldest and most complete
skull of the genus *Homo*, skull 1470, dated to less than 2 million years
old, shows evidence of the Broca's area by a small impression on left
inner surface of the cranium (Leakey and Lewin, 1992: 252).

The secret to our evolutionary success as a species lies in what Ernst
Mayr has called an "open program" (1970: 402–4; 1976: 22–23, 694–
701; 1988: 48–51). Whereas species with closed genetic programs are
restricted in their capacities to adapt to changing environments, depending upon the glacial processes of gene mutations and natural selection,
those species with open genetic programs are able to gather information
not contained in the DNA and to store, recombine, and transmit that data
through cultural evolution. The very existence of an open genetic program suggests, in the words of Konrad Lorenz, that the organism is in
a "state of adaptedness" (1981: 258) in spite of various changes associated with the instability of the environment. Hence, humans, as creatures of open programs, have been able to populate nearly every
latitudinal zone of the Earth's surface, as well as to live outside the
Earth's atmosphere and beneath its oceans.

The development of the open genetic program of human life is most
closely related to the evolution of human intelligence.

SYNAPTIC SOPHISTICATES

While an increase in brain size has certainly contributed to the distinctively human brand of intelligence that has evolved over the past 2.5

million years, the enhanced circuitry that developed *within* the cerebral cortex presumably was of far greater consequence. Studies in neuro-biology have revealed that the significance of any cortical region is related to the internal organization of the synaptic circuits, as well as its connection to other regions of the brain, both cortical and subcortical. In both invertebrates and vertebrates the organization of these synaptic circuits typically follows a kind of polarized, two-dimensional, input-ouput format.

With the extraordinary expansion of the neocortex in humans (the gray matter, which spans the surface of the brain and has more than doubled in the brief transition from *Homo habilis* to *Homo sapiens*), a quantum leap in neurological circuitry took place (Lumsden and Wilson, 1983: 107). Because the human neocortex is a layered structure doubled back on itself, a truly three-dimensional array of circuits was possible, allowing neurons in every layer of the neocortex to be accessible to the inputs and outputs of contiguous layers.

This exceptionally complex and versatile circuitry, combined with the fact that the neocortex is accessible to every major sensory input of the brain, permitted a vast number of ways in which information could be integrated, stored, and recombined. The enhanced flexibility that this multifaceted, three-dimensional circuitry allows is especially important in view of the fact that brain functions do not tend to be localized in certain regions (as was previously thought), but instead are organized in terms of distributed systems involving both hemispheres of the brain (Shepherd, 1988: 623–24, 628).

With increased brain size and cortical sophistication, the genus *Homo* developed markedly enhanced capabilities, in comparison to earlier representatives of the hominid family, to conceive and execute courses of intelligent behavior (for example, "behavior that is adaptively variable within the lifetime of the individual" [Stenhouse, 1974: 31]). This enhanced capacity to behave intelligently reflected the maturation of four primary factors that contribute to the occurrence of intelligent behavior: (1) the P factor—an ability to pause before automatically responding to a situation; (2) the C factor—a memory store for the long-term storage of information; (3) the A factor—the capacity to abstract, to generalize, and to compare and contrast events; and (4) the D factor—the sensori-motor capability to execute a patterned response (1974: 74–128). Sapientization largely describes the maturation of these four factors within humans.

The process of sapientization among humans has opened up new horizons for human thought. Intelligence does not necessarily require thought, but among humans intelligence and thought have become largely synonymous. As intelligent and sentient creatures, finding our place in the world depends first on how we go about thinking of ourselves. Our life consists of more than our thoughts about it, but we cannot understand life apart from our ability to reflect upon it. Hence to know our place in the world is to grasp how we go about representing and understanding ourselves.

THE EVOLUTION OF HUMAN THOUGHT

How did human thought arise? What makes human consciousness different from consciousness in other animals?

In a pathbreaking study of human evolution, entitled *Origins of the Modern Mind* (1991), Merlin Donald charts three stages in the development of human culture and cognition: mimetic consciousness, mythic consciousness, and theoretic consciousness.

Episodic Consciousness

Donald begins by describing a form of consciousness that is evidenced in the behavior of many mammals and birds but finds its fullest development in the apes. He calls it "episodic consciousness." The distinguishing trait of this form of consciousness is its event-bound character. Episodic consciousness is focused on event perception, including the ability to discriminate between certain events—discerning how one incident is different from another—and to store memories about those events. Episodic consciousness is so well developed in apes (but not in monkeys) that it allows for some degree of self-awareness and the limited manipulation of symbols among chimpanzees and gorillas (Calhoun and Thompson, 1988; Gallup, 1970; Gardner and Gardner, 1969; Patterson, 1978; Premack and Premack, 1972).

Take, for example, the case of Kanzi, a male pygmy chimpanzee. Kanzi, born in the Language Research Center of Georgia State University in Atlanta, was adopted by a female chimp named Matata. Matata was the subject of a language study at the center, conducted by Sue Savage-Rumbaugh. The purpose of the study was to determine the ability of the chimpanzee to learn a sign language. After months of work with Matata, the results were disappointing. She simply was not learning it. One day, however, Savage-Rumbaugh noticed something very interesting. Kanzi seemed to be responding to some of the requests and instructions that were directed at his mother in sign language. Savage-Rumbaugh recounts the moment like this ''At first I didn't believe it could be true. But we started to test Kanzi actively, and sure enough, he had learned a lot of words, just picked them up as he played around while we worked with Matata. After that we started to work with Kanzi, and we gave up on his mother.'' Kanzi is now able to respond to involved directives, such as "Go to the bedroom, get the ball, and give it to Rose [a colleague of Savage-Rumbaugh]''—even when another ball is directly in front of him! Kanzi's ability to learn and invent grammatical rules is truly remarkable (Leakey and Lewin, 1992: 243–44).

Mimetic Consciousness

The first stage in the development of human consciousness is the transition from the episodic culture of primate cognition to what Donald calls ''mimetic culture.'' Building upon episodic consciousness, mimetic consciousness introduces gesture-based forms of representation which are consciously self-initiated and intentional yet are not linguistic in orientation.

> Thus, mimesis is fundamentally different from imitation and mimicry in that it involves the *invention* of intentional representations. When there is an audience to interpret the action, mimesis also serves the purpose of social communication. However, mimesis may simply represent the event to oneself, for the purpose of rehearsing and refining a skill: the act itself may be analyzed, reenacted and reanalyzed, that is, represented to oneself. (Donald, 1991: 169)

In mimetic culture the representational manipulation of gesture gradually developed to a degree that permitted toolmaking, the eventual use of fire, big game hunting, and the development of primitive rituals.

Mythic Consciousness

The second stage in the development of human consciousness is the transition from mimetic to mythic culture. This stage is most likely associated with the biological transition from *Homo erectus* to *Homo sapiens* and the descent of the larynx in humans. Mythic consciousness featured the development of semiotic cultures, spoken language, the integration of knowledge, and narrative-based world modeling. Thought, at this mythic stage in the development of human cognition,

> might be regarded as a unified, collectively held system of explanatory and regulatory metaphors. The mind has expanded its reach beyond the episodic perception of events, beyond the mimetic reconstruction of episodes, to a comprehensive modeling of the entire human universe. Causal explanation, prediction, control—myth constitutes an attempt at all three, and every aspect of life is permeated by myth. (1991: 214)

Theoretical Consciousness

The third stage in the evolution of modern thought is the transition from mythic to theoretic culture. This stage of human cognitive development is associated with the diffusion of biologically modern humans (the *Homo sapiens sapiens*) approximately 50,000 years ago. Theoretic consciousness featured the development of external means to store symbols, beginning with the relative explosion of visuographic inventions.

> Unlike body decoration or the purposive arrangement of significant objects, these new graphic inventions were pictorial in nature; that is, they were either two- or three-dimensional representations of recognizable perceptual objects, usually animals. Starting around 40,000 years ago, there was a proliferation of engraved bones and carved ivory: the carvings were highly skilled two- and three-dimensional representations, mostly of contemporary animals. The first evidence of truly advanced painting and drawing skills dates back to about 25,000 years ago, in hundreds of illustrated limestone caves of the Ice Age, of which the best known are at Altamira and Lascaux. Clay sculptures and figurines and a wide variety of trading tokens were common 15,000 years ago. The earliest evidence of writing dates back about 6,000 years, to the emergence of large city-states; and the idea of the phonetic alphabet is less than 4,000 years old. (277–78)

External Symbolic Storage

The cognitive distance from the comparatively recent invention of writing and phonetic alphabets to the development of supercomputers and virtual reality technologies is much shorter than one might think.

The invention of writing represented a quantum leap for the *Homo sapiens* in that it extended the boundaries of normal biological memory by making the external storage and retrieval of information possible.[1]

The development of systems for external memory—what Donald calls "external symbolic storage" (ESS)—can be understood as a kind of culturally channeled "hardware change." It is an integral part of the cognitive architecture of humans, not simply the cumulative consequence of a series of culture-bound "software changes" (308 ff). Formal education itself was invented primarily to enable people to access ESS systems (320). As such, ESS systems opened the way not only for the literal explosion of human cultures across the face of the earth but also for the continual reconfiguration of reality by the manipulation of symbols.

Computer-aided design (CAD) and virtual reality technologies are among the most recent manifestations of the evolution of ESS systems. Most important, unlike the technological developments associated with the mimetic and mythic cultures, the hardware changes associated with theoretic consciousness (i.e., ESS systems) have opened up new pathways for human cognition without any apparent dependence upon genetic change.

> The central point deriving from the history of the third transition, as it moved from visuographic invention to the management of external memory devices to the development and training of metalinguistic skill, is that it was *not* a given of human nature but rather a structure dependent upon both symbolic invention and technological hardware. The hardware may not have been biological, but from the viewpoint of natural history of cognition this does not matter; the ultimate result was an evolutionary transition just as fundamental as those that preceded it. (356)

Could it be that the evolutionary transition to theoretic consciousness also embodies new evolutionary possibilities for human thought? Is the third stage of human consciousness the final stage? Donald seems to leave open the possibility that human thought may not have found its final expression. He concludes his monumental study with these words:

> As we develop new external symbolic configurations and modalities, we reconfigure our own mental architecture in nontrivial ways. The third transition has led to one of the greatest reconfigurations

of cognitive structure in mammalian history, without major genetic change. In principle, this process could continue, and we may not yet have seen the final modular configuration of the modern human mind. Theories of human evolution must be expanded and modified to accommodate this possibility. (382)

THE INTERPLAY BETWEEN DUALISM AND HOLISM

Perhaps the most prominent feature of human cognition is the hybrid texture of the modern mind (Donald, 1991: 355ff.). With tremendous ease, we can shift from theoretic to episodic consciousness, from mimetic to mythic consciousness, and so forth. Human consciousness can be enveloped by the immediacy of sense experience; it can be focused on the reenactment of ritual behaviors; it can soar and recreate reality through narrative imagination; furthermore, human consciousness can be passionately preoccupied with the creation and analysis of external symbol systems.

The interplay between dualism and holism has been of particular significance throughout the evolution of human cognition. On the one hand, dualistic thinking is characterized by the tendency to map conceptual space in terms of opposites or polarities. Holistic thinking, on the other hand, seeks out connections and synthesis within a conceptual space. Both types of thinking are integral features of human thought.

Dualistic thinking most likely lies at the foundations of our capacity for self-awareness. To represent oneself to oneself requires the ability to distinguish the self from the other. Without this self-other polarity as a guide, our conscious experience with the world would take on the character of a blur of sense experience with no discernable point of reference (d'Aquili, 1978; Laughlin and d'Aquili, 1974: 115; Crook, 1980: 243). Moreover, humans seem predisposed to mapping their encounters with reality in terms of polar opposites. Light/dark, good/bad, high/low, hot/cold, and sweet/sour are among the many "binary opposites" that we employ to interpret the world. The significance of this phenomenon of binary opposition has been deftly elaborated by the famous cultural anthropologist Claude Levi-Strauss (1963, 1966). It seems likely that the binary opposites that remain embedded in diverse human cultures ultimately spring from the pivotal self-other duality.

If dualistic thinking resides at the core of our capacity to be self-aware, holistic thinking is a foundational requirement for integrating our knowledge about the world and locating ourselves within it. Presumably, holistic or relational thinking finds its most immediate inspiration from the interconnectedness of biological life. Since the first sparks of human consciousness, humans have been aware of their intimate biological relationship with other life forms. From the depiction of animals in early cave paintings to the development of epic myths, humans have understood themselves as members of a larger biological community—a community characterized by awe and mutual respect.

Whereas dualistic thinking usually places great emphasis on boundary-oriented points of reference, holistic thinking often dismisses such boundaries as arbitrary divisions in its search for connectedness and unity. Mythic consciousness has always provided fertile ground for holistic thinking. While dualistic modes of thinking are always present, to varying degrees, in mythic thought, they are ultimately pressed into service for the "higher calling" of modeling the world—to comprehend expansive landscapes and totalities. This has tended to channel dualistic thinking away from its more extreme and hardened manifestations.[2]

DUALISM AND HOLISM IN THEORETIC CULTURE

Theoretic consciousness, unfettered from the mythic impulse to comprehend totalities, offers ample opportunity for the use and abuse of dualistic thinking. As a consequence, dualistic thinking within *primarily* theoretic modes of consciousness (e.g., Western thought) tends to survey space for the purpose of partitioning reality—drawing boundaries and building walls. The history of the nation-state and colonialism provides stark spacial evidence of dualistic reasoning. The same trend can be seen in the evolution of the concept of private property (e.g., the enclosure movement in England from 1450 to 1640 and from 1750 to 1860; the invention of barbed wire and the "taming" of the American West). At a conceptual level, dualistic thinking has found eloquent expression in pivotal Western thinkers such as René Descartes (humans vs. animals), John Locke (labor-generated value vs. intrinsic value in nature), Immanuel Kant (human rationality vs. sentience/nature), Auguste Comte (science vs. religion), Jean-Jacques Rousseau (humans vs. society), Jeremy Bentham (pain vs. pleasure), and Karl Marx (proletariat vs. bourgeoisie; private property vs. communal property).

Fortunately for our sakes, the most sophisticated recent developments in theoretic consciousness are profoundly destructive of dualistic constructions and wall-building enterprises. From quantum mechanics to evolutionary biology and ecology to astrophysics, we are becoming aware of the profound connectedness of energy and matter throughout the universe. Even the most dualistic machine possible—the binary computer, whose reality is either 0 or 1—has rapidly evolved into a tool that charts relatedness, networks knowledge, and models highly complex systems. These pivotal developments in theoretic consciousness have themselves inspired an entirely new genre of philosophical explorations in holism (e.g., Birch and Cobb, 1981; Capra, 1975; Leopold, 1949; Whitehead, 1929). Moreover, the realities of the Nuclear Age and the Space Age have respectively provided both foreboding and inspiring graphic images of the essential connectedness of life on planet Earth.

It seems quite clear that theoretic consciousness, through the course of its own development, is being infused with new dimensions of holistic thinking. This thinking, representing a creative synthesis between holism and dualism, represents a movement toward seeing reality as an interdependent matrix, where separateness cannot be defined without reference to connectedness. This "interdependent consciousness" not only is an integral part of the modern myths that spark our imaginations and image the future (e.g., *Star Trek*), but also infuses the way in which we understand ourselves as creatures within the biosphere. Ecology has appropriately become the premier science of interdependence.

HOLISM AND HUMAN SEXUALITY

One suspects that the capacity to achieve a satisfying equilibrium between dualistic and holistic thinking (or between skills of discrimination and skills of integration) may have something to do with human sexuality. It appears that women may be generally more predisposed to think in holistic terms—more apt to perceive interconnected networks and to embrace relational thinking. For example, Carol Gilligan's (1982) path-breaking study of moral development among women discovered that women were more likely to understand moral obligation in terms of concrete responsibilities of care within social networks, as opposed to abstract, and often disembodied, principles of justice.

Sara Ruddick (1989) attributes this predisposition toward relational thinking to the profound experiences of child bearing and rearing. Maternal thinking, for Ruddick, is characterized by both concreteness and connectedness. Although women are able to reason abstractly, they tend to reject the demands of abstraction and instead look closely, invent options, refuse closure. They learn to value "connected" ways of conversing:

If concrete cognition does indeed make up one strand of the suppressed and developing different voices attributed to women, we might look to women's maternal work as a partial explanation of this epistemological predilection. It seems a plausible working hypothesis that children's minds would call forth an open-ended, reflective cognitive style in those who try to understand them. A child's acts are irregular, unpredictable, often mysterious. . . . The categories through which a child understands the world are modified as the changing world is creatively apprehended in ways that make sense to the child. If there are comfortable sharp definitions, they are ephemeral. A mother who took one day's conclusions to be permanent or invented sharp distinctions to describe her child's choices would be left floundering. (1989: 95–96)

Along a similar line, the investigations of Mary Field Belenky and her associates (1986) suggest men and women tend to approach the process of knowing in a different fashion. On the one hand, men tend to utilize a procedure of knowing that requires doubt and detachment, as well as the ability to manipulate different conceptual lenses—what Belenky and her associates call ''separate knowing.'' For the most part, formal education is designed to indoctrinate students into the art of separate knowing. One learns to think as an economist, a sociologist, a biologist, or a psychologist. On the other hand, women tend to construct knowledge in ways that require trust and empathy, emphasizing the importance of subjective experience and feeling. Belenkey and her associates call this ''connected knowing.'' Connected knowing thrives upon imaginatively entering another's perspective and abandoning convenient either/or dichotomies. Women who are able to integrate the disparate information they glean from connected knowing

show a high tolerance for internal contradiction and ambiguity. . . . They no longer want to suppress or deny aspects of the self in order to avoid conflict or simplify their lives. . . . They want to avoid what they perceive to be a shortcoming in many men—the tendency to compartmentalize thought and feeling, home and work, self and other. In women there is an impetus to try to deal with life, internal and external, in all its complexity. (1986: 137)

VIRTUAL REALITY AND CONNECTED KNOWING

An important feature of connected knowing is its reliance on empathetic pathways to knowledge. In this regard, it is interesting to speculate on how the development of virtual reality technology could deepen our capacities for empathetic thinking—hence connected knowing. Indeed, virtual reality could become a powerful educational tool to help people to imagine across boundaries, not only deepening their understanding of ecological interdependence but perhaps even revamping the face of theoretic culture as we know it.

For example, what would be the impact on human consciousness if we could experience a range of *virtual habitats*, allowing us to see a habitat from the standpoint of other animals? Think of what it would be like to experience a meadow of alpine wildflowers from the perspective of a bumblebee. Imagine what it would be like to navigate the canopy of a rain forest like a tropical bird. What would it be like to experience the ocean from the standpoint of a whale or dolphin? Or what it would be like to race through a forest like a deer? Would we understand habitats any differently if we could see those spaces from the standpoint of other creatures? Surely the impact of such experiences, for most people, would be profound.

Emergent virtual reality technologies offer tremendous potential to scale and topple the Berlin Walls that obscure our grasp of the wonder of life. When it becomes possible to see and experience virtual habitats through the eyes of other species, it will no doubt dramatically increase our appreciation for the life forms that share the biosphere with us. Could this not also substantially change the way in which humans think? We would understand nature not as an object that we shape and recreate for human purposes but as a community shared by others.

THE MYTH OF OBJECTIVE RE-PRESENTATION

The close association between humans and nature has been embodied in many cultures, including the cultures of Native Americans and other aboriginal peoples (Booth and Jacobs, 1990; Callicott, 1989: 177–219; Naess, 1979) as well as the religious traditions of the East, most notably Buddhism and Taoism (see Callicott and Ames, 1989). Similarly, in early Christian monasticism, there was a strong emphasis on the spiritual kinship of humans and animals (Bratton, 1988). While the emphasis on respect for other life forms has always been a part of the intellectual heritage of the West (e.g., the theme of environmental stewardship in the Bible [Attfield, 1983: 20–50]), this insight has been largely suppressed under the cultural weight of our confusion between representation and objectification.

For human communities, rituals of representation have played a vital role in conveying how we understand the world. Take, for example, the myth of the eternal return in ancient and nonliterate societies (Eliade, 1954, 1959). This widely attested, mythic representation of life is premised on the idea that it is possible to reverse time through the repeated ritual creation of the world. With the reenactment of the creation ritual, life itself is also recreated, and our experience with life becomes cyclical and timeless.

One suspects that our passion with external means of representation—what Donald (1991) refers to as ESS (external symbolic storage) systems—has fostered a myth that is just as pervasive in contemporary literate cultures as the myth of the eternal return was in nonliterate cultures. This modern myth could be termed the "myth of objective re-presentation." This myth embodies the belief that reality, when coded in symbols and reproduced by conceptual structures, assumes a qualitatively new texture. Reality becomes "objective" in the sense that it is re-presented in a way that can be analyzed and manipulated without reference to our subjective relation to reality.

Any venture in symbolic representation runs the risk of confusing the symbol with the thing being symbolized. Yet the habit of re-presenting reality in theoretic culture is so pronounced that our re-creation of the world has diminished the awareness of our essential relatedness to the world. We have sacrificed a connected knowledge of the world in our quest for a separate knowledge of the world. Whereas the ancient myth of the eternal return has transformed the world into a timeless or infinite cycle of life, the modern myth of objective re-presentation has transformed the world into a technologically infinite bundle of resources.

Nowhere is the myth of objective re-presentation more well developed than in mainstream economic theory. In fact, both the neoclassical and Marxist schools of economics are committed to anthropocentric (or human-centered) models of the world that redefine nature as an assortment of natural resources and envision no upper limit on the ability of technology to transform an obviously finite set of resoures into infinite uses.

More generally, the myth of objective re-presentation is plainly evident in our strong faith in technology. It is striking that we are so enamored with our proficiencies as toolmakers that we have made human progress nearly synonymous with technological development. We design societies to fit the latest technology, instead of designing technology to fit our societies. In many respects we have become self-made prisoners of "technological imperatives," believing that everything that can be made should be made (Gabor, 1972).

As such, our faith in technology not only fosters the belief that technological innovations are the answer to nearly every problem, but also reinforces a human-centered approach to the world. The unhappy outcome of this process is that all nonhuman life forms are objectified and ecosystems are valued according to the extent that they contribute to human purposes.

NATURE IN NATIVE AMERICAN PERSPECTIVE

A stark example of the myth of objective re-presentation can be seen in the behavior of European settlers as they "tamed" the American West. Their behavior is all the more unsettling in light of the sharp contrast between their actions and attitudes and the way in which the indigenous peoples of America understood their relationship to the environment.

Civilization, for the European settlers, was understood as the process of subduing and controlling the forces of nature—including, of course, the indigenous peoples. The wilderness was "fenced out" by barbed wire, the stage coach, and rail transportation. Small communities brought their own varieties of law and order to the wilderness. Eventually land was cleared (i.e., destroyed) for "productive" uses. From the standpoint of the settlers, land was reduced to a mere resource that could be owned and partitioned. Both the buffalo and the Indian were objectified and "cleared" by the rifle and the Colt six-shooter. Nature, for settlers of European heritage, was an object to be transformed, and the destruction of wilderness was noble, their "manifest destiny."

In the minds of the Native American peoples, there was nothing civilized or noble in the actions of the settlers. They came as arrogant marauders, felling buffalo for their skins or for sport, leaving their carcasses to rot on the plains. By contrast, the native American tribes understood themselves as being part of a larger community of creatures. The buffalo was not a commodity to be exploited but a member of the Native American's community. As Calvin Martin writes;

When Indians referred to other animal species as "people"—just a different sort of person from man—they were not being quaint. Nature was a community of such "people"—"people" for whom man had a great deal of genuine regard and with whom he had a contractual relationship to protect one another's interests and fulfill mutual needs. (1978: 187)

Moreover, nature, from the perspective of the Native Americans, was coextensive with life itself: The earth, the water, the rocks and winds all have life. Luther Standing Bear, a Lakota Sioux, put it in this way:

There was no such thing as emptiness in the world. Even in the sky there were no vacant places. Everywhere there was life, visible and invisible, and every object possessed something that would be good for us to have also—even to the very stones. . . . Even without human companionship one was never alone. The world teemed with life and wisdom; there was no complete solitude for the Lakota. (1933: 14)

For Standing Bear the recognition of our connectedness with nature provokes a profound respect and love for nature.

We are of the soil and the soil is of us. We love the birds and beasts that grew with us on this soil. They drank the same water as we did and breathed the same air. We are all one in nature. Believing so, there was in our hearts a great peace and a welling kindness for all living, growing things. (45)

The holism that pervades the worldviews of Native American tribes is striking to Western eyes; the human being reverences nature and lives within it, not over it. The Native American's all-encompassing, holistic perspective stands in stark contrast to Western worldviews which have been strongly influenced by both religious and scientific varieties of dualistic thought. While it is difficult to know how the ecological wisdom of Native American peoples can be translated into Western worldviews, it may be more significant for our purposes to speculate on what might lie behind the environmental sensitivity of Native American peoples.

LESSONS FROM THE LATE PLEISTOCENE?

We know that the indigenous peoples of the Americas were the direct descendants of the Siberian big game hunters who crossed the Bering land bridge over 11,000 years ago. We also know that, during this same period, the late Pleistocene, there were massive extinctions of over 30 genera of megafauna (large mammals weighing more than 100 pounds).

Was it mere coincidence that the arrival of big game hunters from Siberia was marked by the massive extinction of mammalian species in the North American continent? While one cannot say for sure, it may be that these Siberian hunters had something to do with the extinction. If this was the case, one is tempted to understand the ecological wisdom of Native American peoples as an evolutionarily adaptive response to the abrupt extinction of large mammals across the North American landscape. As Paul Martin suggests:

> With the extinction of all but the smaller, solitary, and cryptic species, such as most cervids, it seems likely that a more normal predator-prey relationship would be established. Major cultural changes would begin. Not until the prey populations were extinct would the hunters be forced, by necessity, to learn more botany. (1973: 972)

If the ecological sensitivity of Native Americans reflects, in part, the remnants of an archaic adaptation by the first inhabitants of North America, then one can hope that humans of the late twentieth and early twenty-first centuries will craft similar adapative responses to the massive extinctions and ecological deterioration that characterize our day, transforming our parasitic relationship to the environment into a mutualistic relationship.[3] One suspects, though, that our adaptive response will rely less upon innovations related to our mythic human consciousness and more upon emergent developments in theoretic human consciousness. In this regard, it seems likely that the alliance between myths of New Age genre and segments of the environmental movement have tended to weaken the credibility of environmental concerns in many political arenas.

One creative attempt to blend aspects of human mythic culture and theoretic culture in response to the present ecological crisis is found in James Lovelock's Gaia hypothesis (Lovelock and Epton, 1975; Lovelock, 1979, 1990).[4] Utilizing the term "Gaia," the nurturing earth goddess in ancient Greek mythology, Lovelock argues on the grounds of biochemistry that we need to understand the planet and its atmosphere as a living organism, whose well-being supersedes the well-being of any individual parts (e.g., humans). As such, Earth, or Gaia, as a living entity, has a moral status in the same way that people have certain rights to be protected and sustained. Hence, Lovelock combined the resources of Greek mythology (mythic consiousness) and modern biochemistry (theoretic consciousness) to articulate a truly biocentric understanding of ethical obligation. One can only hope that we will see many more creative efforts in revitalizing themes that have been a part of our mythic heritage by placing them on a solid theoretic foundation.

The myth of objective re-presentation has yielded a lifeless world that offers neither connection nor communion. The myth could not exist if it were not for the Berlin Wall etched in our minds, a wall that isolates us from the larger community of the life that surrounds us. It is difficult to live without Berlin Walls. We have become so accustomed to our self-appointed, cherished status as "intelligent" life forms—God's gift to the planet. Letting go of our delusions of grandeur will not be easy. Yet it is only in letting go that we will be able to find ourselves—to find our place in the world and to rediscover the strength of our humanity.

NOTES

1. This is not to minimize, though, the significance of the technological and sociological innovations that preceded writing (e.g., the invention of agriculture, the development of fired ceramics, percussive musical instruments, the development of lunar records [notched bones], the first maps, the invention of sailing craft, the development of towns and cities). Donald emphasizes this point when he writes:

The complex technological and social developments that preceded writing might suggest the existence of some apparently analytic thought skills that contained germinal elements leading to later theoretic development. However, early inventions were pragmatic and generally not far removed from nature: for example, the domestication of animals and plants would not have required more than a recognition, transmitted over time, that certain species were desirable and domesticable for human use. Complex constructional products, such as brick structures and sailing vessels, might be seen as grand elaborations on the ancient toolmaking skills of humans. The social organization of the first towns and cities presumably borrowed heavily from existing family and tribal structures. These pragmatic developments, impressive as they were, lacked the essentially reflective and representational nature of theory. (1991: 334–35)

2. For example, the thoroughgoing dualism of ancient Zoroastrianism was tempered by the belief that Zoroaster's great creator god, Aura Mazda, meaning "Wise Lord," was the progenitor of the "twins" Spenta Mainyu (Holy Spirit) and Angra Mainyu (Destructive or Evil Spirit). Moreover, Aura Mazda "stands above and beyond" the perpetual and cosmic conflict between these two spirits (see Zaehner, 1961: 43, 50–52).

3. It is likely that most forms of symbiotic mutualism evolved from originally parasitic relationships (Ahmadjian and Paracer, 1986: 4).

4. See Nash for a discussion on the influence of the Gaia hypothesis for contemporary environmental thought (1989: 157–59).

Chapter 3

Specious Speciesism in Ethics and Economics

SPECIOUS SPECIESISM AND COGNITION CONCEIT

Sum, ergo cogito—I am, therefore I think.

The simple transposition of René Descartes' *cogito, ergo sum* places human cognition in its proper perspective. We are a wondrous species who have been given the power of rational thought—a miraculous gift of our evolutionary history. It is difficult to fathom the tremendous opportunities and responsibilities we have by virtue of being thinking creatures. We do not need to build a Berlin Wall between humans and all other species in order to appreciate the uniqueness of our life form. We *are* unique. Humans have no need to prop up their creaturely status by placing themselves at the center of the universe. "Being human" is not a synonym for weakness, but a source of strength. Arrogance only detracts from that strength.

Cognition conceit, like the conceit of a playground bully, is a sign of weakness. It is the cornerstone of the Berlin Wall between humans and all other life forms. In our weakness, we have degraded the tremendous asset of human rationality by turning it into a rationale for domination. Humans characteristically define intelligence in a manner that favors our inherent strengths as toolmakers. But what of the intelligence of a migratory bird that uses the stars to navigate across thousands of miles every year? How do we account for the intelligence of a whale that repeats, without flaw, the complex and lengthy series of tonal variations in a mating call, or of the intelligence of a salmon who returns to a tiny tributary after years of navigating vast reaches of oceans?

We have conveniently grown used to dismissing these forms of intelligence as instinct. It is as if we are not satisfied with trumping up our peculiar genre of intelligence. We must also discredit the intelligent behavior evidenced by other members of the animal kingdom. Indeed, it is appropriate to ask, what would it take before we recognized intelligence in other species? Would they have to be toolmaking creatures like us? Must they be able to speak?

THE LONELY HUMAN

Speciesism is racism by another name. The driving forces that have fostered racist attitudes toward other humans are the same rationalizations that have accelerated the spread of earth cancer. Just as the racist is a prisoner of his or her own hatred, so are we prisoners of our own ecological isolation—alienated within our own habitats. When we have the opportunity to experience the wilderness and the beauty of the natural world, it reminds us of how impoverished we have become. Cognition conceit has prevented us from seeing nature as *our* community. How else could we live with ourselves as cancerous colonizers?

The manifest lack of being in community with the natural world is apparent in almost every facet of contemporary culture. For most people, the only meaningful relationship with this ecological community comes by spending a summer vacation in the wilderness or by interacting with a house pet. Yet, despite the brief and sporadic nature of these encounters, they usually have a profound impact on us.

Who, for example, after experiencing the glory of a national park would want to parcel it up and sell it for private development? Certainly some land developers would approve, but not those who experience the majesty of the habitat as a whole. We ascribe a certain sacredness to untouched wilderness; the aesthetic quality of natural beauty is in a fundamentally different category than manmade beauty.

The extent to which we commune with nature will decisively impact the degree to which we perceive the natural world as our community. Just as Berlin Walls obscure our recognition of ecological community, the absence of Berlin Walls understandably deepens our sense of oneness with the natural world. Nowhere is this more plainly apparent than in the way in which children see the world around them. For a child, the natural world is a playground of unfathomable mystery. Growing up, unfortunately, is usually synonymous with socialization processes that replace a child's sense of wonder with matter-of-fact, bite-sized portions of knowledge. Instead of extending the marvel of childhood into adulthood, adulthood carries with it a tacit responsibility to deny "childish" ways.

Children have much to teach adults about their relationship to the natural world. It is not surprising that children respond passionately to ecological education—typically goading their parents into practicing environmental responsibility—in light of their sense of connection to the natural world. From well-worn fairy tales to contemporary cartoons, animals are full participants within the human community—either interacting with humans on an equal basis or modeling essentially human traits and situations within the exclusive domain of the animal kingdom. One cannot browse through the children's book section of any bookstore without being impressed by the profound role that animals play in a child's imagination. No other genre of literature is more "animalcentric" than that of children's literature.[1]

Take, for example, a children's book, entitled *The Best Nest*, by P. D. Eastman. The book uses the life story of Mr. and Mrs. Bird, with a hint of sexism, to illustrate the value of appreciating one's home. The story opens with Mr. Bird happily singing a tune that celebrates the beauty of their home. Mrs. Bird, upon hearing the tune, protests her husband's cheery assessment of their home and demands to relocate their nest. As the story unfolds, Mr. and Mrs. Bird consider a number of alternatives for their new home, none of which is satisfactory. Finally Mr. and Mrs. Bird elect to build their nest within the belfry of a church. This was, according to Mrs. Bird, the best nest. Unfortunately for them, however, the nest was unwisely constructed on top of the bell itself. When the bell rang, Mrs. Bird fled the belfry in terror. She returned promptly to their old nest. Mr. Bird returned to the belfry and found the nest in ruins; Mrs. Bird was nowhere in sight. Frantic, he searched high and low for Mrs. Bird to no avail. Finally, in desperation, he returned to their old nest only to find Mrs. Bird happily singing the same joyful tune that her husband had sung.

Inevitably, the imaginative musings of children give way to adult concepts of the world. Mystery gives way to explanation, and knowledge supplants ignorance. Unhappily, though, our socialization into adulthood all too often empties the world of wonder. There is nothing inevitable about this process; it is the educational by-product of the Berlin Wall that separates humans from all other life forms. Authentic knowledge about the natural world cannot help but provoke awe and marvel; it opens horizons instead of restricting them. The "lonely human" has yet to find his place in the world, being trapped in a world that retains the semblance of manageability while being much too small to be captivating.

I THINK, THEREFORE I HAVE PURPOSE

One primary consequence of cognition conceit is the habit of disregarding any purposeful dimension of animal life. We often assume that purpose—or purposeful living—cannot exist apart from rational forms of consciousness. Immanuel Kant himself lectured his students that "so far as animals are concerned, we have no direct duties. Animals are not self-conscious, and they are merely as a means to an end. That end is man."[2] Since, for most of us, our immediate experience with animal life comes in the form of domesticated house pets, this tends to reinforce the perception that animals simply exist—they eat, yawn, and sleep—and have no real purpose in life.

The modern-day institution of factory farming is certainly one of the more distressing manifestations of our disregard of purpose within the animal kingdom. If we genuinely believed that animals had purposes of their own, we simply could not justify the way we cultivate animals for our own consumption—raising them in dark, cramped cages with wire floors or stalls that are so small that an animal cannot move. If we did ascribe purposeful living to animals, it would radically change our perspective. Certainly our perceptions of animal purpose would be greatly altered if we observed firsthand the struggles of animals in their natural habitats—if we grasped something of the labor associated with bearing and protecting the young, building nests, finding food, and fending off predators.

It is curious that humans would go out of their way to define purpose in terms of human rationality and then use purpose as a barometer for the worth of life. Most of us have great difficulty in finding purpose in life! Many people feel compelled to go *beyond* life—to either "see behind" life or transcend life—in order to find purpose in life. This may partly reflect our profound alienation from nature. For "civilized" human beings, our estrangement from nature is so acute that we are fastidiously preoccupied with the fundamental biological processes of bodies (e.g., eating, defecation, sex, aging)—showing particular distaste for bodily fluids. In the Bible, the first evidence of human alienation, after eating from the "tree of the knowledge of good and evil," was Adam and Eve's need to cover their bodies (Gen. 3:7). Wholeness for humans requires an understanding of our essential connectedness to all life forms that share in our planet's bounty.

There is nothing "intelligent" about cognition conceit. Truly intelligent (i.e., adaptive) behavior for humans at the end of the twentieth century would call for a more balanced and less arrogant self-understanding. Like the Siberian hunters who first discovered the Americas over 11,000 years ago, we must learn to live in an interdependent relationship with our environment, not a cancerous one. Unfortunately, cognition conceit still enjoys an axiomatic status in two critical regions of human knowledge: ethics and economics. Leading representatives of both fields of knowledge are true believers in the myth of objective representation and cling tenaciously to the cultural foundations of cognition conceit.

Curiously, both ethics and economics share a contradiction that is alternatively humorous and macabre. Both disciplines share a strong bias against the use of coercion as a strategy for control and colonization. In contemporary ethics, this bias is reflected in a strong commitment to individual autonomy. In economics, the preference against coercion is most clearly expressed as a bias against monopolistic modes of production and allocation. Yet, despite these value commitments, both economics and ethics have legitimated—even celebrated—the most grotesque genre of coercive and cancerous colonization known to our planet's biological history. This contradiction is no paradoxical accident of history. It is only made possible because of Berlin Walls. In the case of ethical theory, it is a Berlin Wall between morality and creativity. For economics, a Berlin Wall divides household from habitat. The remainder of this chapter, along with the next two chapters, is devoted to examining this contradiction more fully.

STRANGE BEDFELLOWS: ETHICS AND ECONOMICS

Ethics and economics share an interesting history. The father of economics, Adam Smith, was also the first person to hold a formal academic chair in moral philosophy. Moreover, John Stuart Mill, a pivotal figure in both moral and political philosophy, was also an an influential economic thinker in his own right. Mill's *Principles of Political Economy* (1848) was considered the foremost text in economics until the publication of Alfred Marshall's *Principles of Economics* in 1890. In addition, prior to Smith, some of the most influential ethical thinkers also contributed substantially to the pre-Smithean body of economic knowlege (e.g., Plato, Aristotle, and Saint Thomas Aquinas).[3]

What is especially noteworthy about the developments in ethics and economics since the eighteenth century is the central place that human rationality occupies in the methodologies of both disciplines. In ethics, the influence of Kant has been immense. He is a watershed intellectual figure for contemporary ethical theory in much the same way that Smith is for economic theory. Although Smith and Kant constructed their theories on the basis of different notions of human rationality (with Kant's being the most adequate philosophically), both men shared a considerable confidence in human rationality and its universal character. Smith's understanding of market-based economics was strongly influenced by his moral reevaluation of self-interest, his confidence in instrumental rationality (i.e, the ability to choose means appropriate to a specified end), and his belief in universal moral sentiments.[4] Kant's understanding of ethical obligation was informed by his confidence in "practical reason," his belief that human rationality is both substantive and universal, and his conviction that an action has moral worth only if it is divorced from natural inclinations.[5]

Underlying the respect for human rationality shared by Smith and Kant is the conviction that human communities—and, indeed, the entire planet—should be ordered according to the precepts of human rationality. This belief, of course, is central to all Enlightenment thought, of which Kant and Smith were able representatives (along with Karl Marx as a post-Enlightenment, Enlightenment thinker).

During the twentieth century, developments in both ethics and economics reinforced this fundamental confidence in human rationality as a principle for ordering human communities. Even the nihilistic movements of emotivism and existentialism evidenced this confidence in human rationality.[6] In neoclassical economic theory, the instrumental rationality of Economic Man (*Homo oeconomicus*) laid the conceptual foundations for welfare economics and opened up new territory for marginal analysis.

While ethics and economics began to part ways with the enunciation of the concept of a value-free, positive economics in Nassau William Senior's *An Outline of the Science of Political Economy* (1836), the divorce was not final until the demise of the notion of cardinal utility at the hands of British economist Lionel Robbins (1932). From that point on, economics became the science of optimizing subjective preferences, and many embraced the belief that it was possible to do economic analysis without making judgments of value. This assumption of a value-free approach to economic analysis issued in what Gunnar Myrdal called "a perpetual game of hide and seek in economics" consists of "hiding the norm in the concept" (1953: 192). Although this notion of a value-free approach to economic analysis has been roundly criticized (Balogh, 1982; Hirsh, 1976; 137–51; Katouzian, 1980: 45ff; Neill, 1978; Sen, 1977; Thurow, 1984), the belief is alive and well in classrooms, economic textbooks, and university seminar rooms around the world.

Since the split between economics and ethics, some observers have felt that fundamental economic concepts needed to be explicitly reformulated with respect to ethical theory and other social sciences (Etzioni, 1988; Sen, 1987). The process envisioned usually amounts to refashioning the connection between ethics and economics. Reconnecting these fields would presumably require abandoning the assumption of value-free economic analysis, introducing ethical concerns directly into economic analysis. To date, the boldest thinking associated with the reintegration of ethics and economics has related to the ecological sphere (Daly, 1991; Daly and Cobb, 1989). The question must be asked, though, whether these well-intentioned attempts at integration are flawed, to a fatal degree, by the underdevelopment of ethical theory?

ETHICS AND ECONOMICS—YE MUST BE BORN AGAIN

When one looks at both ethics and economics from the standpoint of ecological interdependence, both fields are fraught with substantial methodological deficiencies. While the environmental ethics literature has blossomed significantly within the past decade,[7] it still suffers from a "sideshow" status in relation to mainstream ethics. Certainly the core insights emerging from this literature have the power to revolutionize ethical concepts from their very foundations (Spitler, 1985). Yet most environmental ethicists have treated their subject matter more as an ethical excursion into ecological issues than as an opportunity for a radical reformation within mainstream ethics (with Callicott, 1980, 1987, 1989; Johnson, 1991; Rolston, 1986, 1988, 1991 as notable exceptions).

The "sideshow" status of environmental ethics also characterizes environmental economics and the emergent field of ecological economics. While environmental economics has achieved an aura of respectability within the discipline, due primarily to the methodological respectability of "environmental impact analysis," it has thus far demonstrated little concern for critiquing the paradigm assumptions of mainstream economic thought. Environmental economics, like the environmental movement in industrialized countries, has been preoccupied primarily with issues of pollution, and profound ecological issues have been trivialized as "externalities" (MacNeill, Winsemius, and Yakushiji, 1991). In fact, when one looks critically at how "environmental protection" is understood by most governments (i.e., cleaning up the industrial wastes that threaten human habitats), a good argument can be made that environmental economics may do more harm than good by legitimating the status quo—creating the illusion of ecological concern while leaving perniciously parasitic habits of resource utilization unchallenged. While the emergent field of ecological economics (Costanza, 1991) has considerably more potential to critique the governing assumptions of mainstream economic theory, the field has not yet been accorded the attention it deserves by mainstream economists.

Both ethics and economics have suffered greatly under the weight of their pre-ecological disciplinary heritage. It was not an ethicist but a naturalist, Aldo Leopold (1933, 1949), who effectively launched the field of environmental ethics. Moreover, it seems reasonable to assume that environmental economics would not exist today if it had not been for the political pressure exerted by the environmental movement (largely influenced by biologists and ecologists).

The capacity of ethics and economics to reform themselves has been severely hampered by their commitment to dualistic notions of the universe and their enthusiastic endorsement of the myth of objective representation. To no small degree, both disciplines have tried to remake the world according to their own image (i.e., after their particular understandings of rationality).

In economics, the project of remaking the world is epitomized by the cherished yet outdated notion of Economic Man (*Homo oeconomicus*). This conception of human nature understands people as utility-maximizing and egoistic creatures whose wants are insatiable. Economic Man, combined with a Ricardian penchant for abstraction and deductive reasoning, explains much of what is wrong with economic theory today. One could not have conceived a better justification for virulent and self-destructive cancerous colonization than that of *Homo oeconomicus*. Despite the mounting and persuasive evidence against this concept of human nature (Collard, 1978; Etzioni, 1988; Lea, Tarpy, and Webley, 1987), it still remains a key conceptual peg of microeconomic theory.

For ethical theory, the fundamental understandings of value and obligation are rooted in the pervasive belief that humans remain at the center of the moral universe. Because human rationality is conceived as the basis of moral value, after Plato, Aristotle and Kant, moral obligation is either something only shared among humans for the sake of humans or something that humans may generously elect to extend unilaterally to other creatures (of course, always on their own terms). This anthropocentric assertion is endemic to the entire ethical enterprise.

When ethics and economics are denied their familiar anthropocentric moorings, both fields look more like fantastic delusions than serious academic disciplines. The pervasive myth of objective re-presentation leads economists to objectify all that is not human and to order all things according to human preferences—reducing them to mere ''consumption bundles'' and ''factors of production.'' It encourages ethicists to assume that the moral universe is restricted to a small portion of the biosphere, where genuine moral value issues from the purposes and projects of one species.

What justification can be found—rational or otherwise—for the way in which mainstream economic theory transforms the biosphere into a bundle of natural resources for the insatiable desires of *Homo oeconomicus*? What warrants the judgment of mainstream ethical theory that categories of moral value apply only to one species in a biosphere teeming with life? Why would two disciplines that otherwise despise the use of coercion as a strategy for control and colonization embrace and encourage earth cancer? Are these the conclusions of human rationality in the late twentieth century? Or do they result from the blindness to which we are accustomed . . . living in the shadow of the Berlin Wall?

INTERDEPENDENCE: FROM ECOLOGY TO ETHICS TO ECONOMICS

Ecology, as the premier science of interdependence, has the conceptual power to rescue both ethics and economics from the morass of self-defeating anthropocentrism. Its robust concept of interdependence not only has the ability to link these disciplines, but also can provide ethics and economics with a firm theoretical foundation.

Interdependence, as the term is currently used in ethics and economics, usually finds a restrictive application in connection with interhuman or interstate relationships. Most often, the term finds application in global contexts involving human-human relationships (e.g., North-South interdependence, the interdependent global economy). Significantly, the authors of a follow-up report to the pathbreaking work of the Brundtland

Commision (*Our Common Future*, 1987) felt constrained to entitle the work: *Beyond Interdependence: The Meshing of the World's Economy and the Earth's Ecology* (MacNeill, Winsemius, and Yakushiji, 1991). Even the highly progressive report of the South Commission (1990), chaired by Julius Nyerere and entitled *The Challenge to the South*, defined global interdependence in such a way that the biosphere becomes a backdrop (or enforcer) of interdependence among humans, instead of extending the concept to human-biosphere relations. In the final chapter of the report, we read the following under the heading of "The South and the Management of Global Interdependence":

> The concept of global interdependence describes a fundamental trend in the modern world. The interrelationships among countries have multiplied and diversified to an unprecedented degree. International flows and transactions tightly enmesh all national economies, while transport, communication, and information networks span the globe. The biosphere reacts globally to man's intrusions irrespective of where they originate. (1990: 283)

The fact that the report devotes only two sentences (pp. 135–36) to the loss of biodiversity—noting the adverse consequences of species extinction for not only "ecological imbalances" and "biological diversity" but also "the production of valuable industrial and pharmaceutical substances"—is a profound indication of the woeful inadequacy of definitions of interdependence in current parlance.

With the exception of the growing environmental ethics literature and the emergent field of ecological economics (Costanza, 1991), neither mainstream ethics nor economics has been concerned with interdependent relations between humans and the biosphere. Presumably this is largely due to the influence of the myth of objective re-presentation in both disciplines. One does not have an "interdependent relationship" between subjects and objects, only subjects and subjects. If the biosphere is simply an objective backdrop for human projects and purposes, one cannot have a meaningful concept interdependence in human-biosphere relations. Sadly, the current notions of interdependence are steeped in the mythology of objective re-presentation.

Fortunately ecology, and the life sciences in general, are free from the restrictive and imperialistic notions of interdependence that have been so influential in ethics and economics. For ecologists, interdependence characterizes the systemic and symbiotic relationship of all life forms in an ecosystem. No creature can exempt itself from the essential interdependence of the biosphere. The "deep structure" of all of Earth's habitats proclaims the profound interconnectedness of life. To deny this is to deny our biological existence.

Ecology holds within its disciplinary purview the insights that have the capability to refashion both ethics and economics and to demythologize the truncated concepts of interdependence that issue from the social sciences. One suspects that the "line of transmission" among ecology, ethics, and economics would begin with ecology, proceed to ethics, and gradually make inroads into economic theory. Given the dependence of contemporary economic analysis on "covert" ethical judgments, it is difficult to conceive of a meaningful dialogue between ecology and economics if ethical theory is relegated to the sidelines. Moreover, at least one connection between two of these disciplines (ecology and ethics) is being solidly forged in the maturing environmental ethics literature.

To synthesize insights from ecology, ethics, and economics within a single conceptual framework makes considerable sense. Ecologists and other biologists, for example, need to become increasingly literate in ethics and economics in order to execute effectively their obligations as citizens to press for environmental responsibility in the public sphere. The separation between good science and good citizenship is no more acceptable in biology than in nuclear physics.[8] Accordingly, if ecologists become involved in environmental education, they should acquire a baseline knowledge of ethical theory and moral development. Moreover, ecologists and other biologists must necessarily make connections between ecology and economics when they articulate environmental concerns in connection with public policy.

Similarly, the interdependence of these three disciplines is strongly evident in the case of ethical theory. It is difficult to envision the development of robust ethical theories apart from substantial contact with the disciplines of ecology and economics. Ecology rescues ethics from its self-defeating heritage of anthropocentrism, and economic analysis provides ethicists with a tool kit and vocabulary to address ethical concerns, in particular policy environments.

Unfortunately, the distinction between "pure" and "applied" ethics has not served the discipline well; it has diminished the capacity of ethical theory to renew itself in dialogue between other disciplines. The trajectory of applied ethics is usually from moral philosophy to a relevant social science or life science field without substantial feedback loops back to moral philosophy. Rarely does this interaction produce fundamental changes in how ethical theory itself is conceived. Accordingly, ethicists have generally confined themselves to "small conversations" with philosophy and theology when it comes to formulating ethical theory, typically only broadening these conversations when it comes to applying the theory in real world contexts.[9]

Finally, economic theory stands in desperate need of revitalization. Although economics has the reputation of being the most scientific of the social sciences (e.g., Nobel prizes are awarded in economics but not in political science, sociology, anthropology, etc.), the discipline has traded practical relevance for theoretical elegance. Unlike the expansive and eclectic thought of such pioneering economic thinkers as Adam Smith, John Stuart Mill, Alfred Marshall, and John Maynard Keynes, contemporary economics is characterized by a compulsive drive for specialization, confusing truth with methodological rigor.

The credibility gap between economic theory and economic reality is increasing at a dramatic pace. Within the field, this credibility gap is most evident in the breakdown of macroeconomic theory—most likely the result of inadequacies in microeconomic theory (see Thurow, 1984). Moreover, critics both inside and outside the discipline have long noted the tendency in mainstream or neoclassical economics to disregard historical, institutional, cultural, and ecological factors in understanding economic realities. Scientific objectivity has been confused with the decision to see narrowly.

Modern life is laced with self-defeating and imprisoning forms of specious speciesism—nourished by self-serving definitions of intelligence and fed by delusions of grandeur. In reality, it is humans, not animals, who stand in need of liberation. Our redemption and reconciliation with planetary life requires us to celebrate the essential interdependence of life, practicing humility and respect toward all creatures. Our liberation is Earth's liberation from cancer's clutch.

NOTES

1. It could be argued, of course, that the presence of animals in children's literature is more a reflection of a child's imaginative capacity for anthropomorphism than any predisposition toward biocentrism. Yet, even if this is the case, the habit of anthropomorphism must surely enhance a child's capacity to identify with other creatures, thereby deepening the impression that both humans and animals share common interests. In adulthood, a similar anthropomorphic impulse is typically evidenced by the owners of household pets.

2. From Kant's *Lectures on Ethics*, trans. by L. Infield (New York: Harper Torchbooks, 1963), pp. 239–40.

3. Of special significance was the way in which Saint Thomas' analysis of a "just price" recognized the influential role of supply and demand in price determination. See Worland (1967).

4. Smith's *An Inquiry into the Nature and Causes of the Wealth of Nations* (1776) is often read and interpreted without reference to his *The Theory of Moral Sentiments* (1759). As a consequence, there has been a tendency in the history of economic thought to assume that the market itself will function effectively because of its own benevolent rationality, instead of building upon the common moral sentiments of humanity.

5. For example, in the *Groundwork of the Metaphysics of Morals* (1785), Kant argues concerning an action that is not motivated by duty or reverence for the moral law (i.e., the categorical imperative):

To help others where one can is a duty, and besides this there are many spirits of so sympathetic a temper that, without any further motive of vanity or self-interest, they find an inner pleasure in spreading happiness around them and can take delight in the contentment of others as their own work. Yet I maintain that in such a case an action of this kind, however right and however amiable it may be, has still no genuinely moral worth. It stands on the same footing as other inclinations. (10)

6. For emotivism, this confidence in human rationality was expressed by proponents of emotivism who held that once moral concepts were clarified through linguistic analysis (i.e., demonstrated to be meaningless), we could seek solutions to the real problems that divide people. With respect to existentialism, this confidence in rationality was present in the belief that one could shed self-deception and live truly authentically with freely chosen, self-generated modes of meaning.

7. The most important academic journal in this field is *Environmental Ethics*, which began in 1979. See Nash (1989) for an excellent history of the movement.

8. It seems that ecologists and naturalists have grasped this insight more than those in other scientific disciplines. For example, the international community is greatly indebted to a rugged, highly devoted cadre of scientists who endure harsh (and sometimes life-threatening) circumstances to collect data on biodiversity and environmental degradation. Moreover, ecologists and naturalists have been at the forefront of the environmental movements of industrialized countries around the world.

9. Of course, there are a few notable exceptions to this generalization. For example, Rawls (1971) appropriated insights from game theory (a field most closely related to economics and political science) in his argument for the two principles of justice from the "original position," even though Rawls' "maximin" decision rule contradicts the extant evidence from game theory (see Arrow, 1973; Barry, 1973: 87–107).

Chapter 4
The Ethics of Earth Cancer

THE ANTHROPOCENTRIC FALLACY

Ethics, a discipline with a long and distinguished history, has been hampered by what I will refer to as the "anthropocentric fallacy." This fallacy holds that because humans have the capacity for ethical reflection and moral decision making, humans are also the proper ends (usually the exclusive and sole ends) of morality. As such, the anthropocentric fallacy confuses the human capacity for ethical reflection with the aims of ethical theory. Those who cannot enter the dialogue about moral values are definitionally excluded from the protective umbrella of morality. The fallacy arises from a restrictive qualification in the conclusion that is not contained in the premises. This "fallacy of accent" is as follows:

Premise 1: Human beings are moral decision makers.
Premise 2: Moral decision makers are subjects of morality.

Conclusion: Human beings are *the* subjects of morality.

As I will argue later in this chapter, the anthropocentric fallacy could not maintain itself if it were not for the Berlin Wall that exists between morality and creativity, a divide that effectively disqualifies creativity as an essentially moral category. As such, morality is seen as a constraint on creativity, instead of creativity being understood as a moral attribute. This Berlin Wall has been sustained by the artificial distinction that philosophers have made between the moral and aesthetic dimensions of life.

From its beginnings, the field of ethics has evidenced a pronounced anthropocentric bias. The Greek word, *ethnos*, from which the term "ethics" is derived, properly refers to the customs of human communities. Hence the study of ethics, in the original sense of the term, was understood as the study of the customs, habits, or mores of human communities. Even highly progressive, "expanding circle" (Singer, 1981) understandings of moral obligation still place humans at the moral center of their theories.

THE ANTHROPOCENTRIC FALLACY IN PLATO AND ARISTOTLE

In Plato's *Republic* the anthropocentric bias of ethical reasoning is most clearly evident. Plato defined justice in terms of what the ideal state would look like. He constructed that state deductively from his understanding of the human soul, which he understood to consist of three elements: (1) an appetitive part, (2) a spirited part, and (3) a rational part. Plato argued that the rational part of the human personality was superior to the spirited part, and the spirited superior to the appetitive, and he proceeded to create an ideal society "in speech" (i.e., in theory) that was modeled on this understanding of human nature. Hence, the concept of justice, for Plato, is defined by the just state, and this society is necessarily based on the core components of the human person, as he understood them. Plato's *Republic* is the human soul writ large.

The anthropocentric bias in Aristotle is less blatant but considerably more influential in the history of ethics. Using an inductive methodology, Aristotle, in his famous *Nicomachean Ethics*, defines good ethics in terms of the way in which good *men* lived. Qualities such as courage, generosity, prudence, and the like are virtues because they are manifest in the lives of good men (who also live in moderation and subscribe to the "golden mean"). Happiness or *eudaimonia* (i.e., well-being) was understood as the end toward which good men order their lives. Because humans are endowed with the faculty of reason, humans have the potential to cultivate the habits of virtue and to live with a profound sense of well-being. Although Aristotle's *Ethics* begs some important questions (e.g., Is the definition of a good person relative from one culture to the next?) and excludes the possibility that the actions and character of good women also contribute to the portrait of a virtuous life, the influence of his work in the history of ethics is immense.

Except perhaps for the writings of Immanuel Kant, Aristotle still has no equal in the history of ethics in the West. In Aristotle's thought the anthropocentric fallacy became firmly rooted in the ethical enterprise. Whereas the anthropocentric commitments of Plato's theory were overshadowed by his crude anthropology and excessive features associated with his ideal state, Aristotle's *Ethics* provided the definitive geography for subsequent ethical thinking. This geography located *human* well-being at the very center of the ethical landscape. All other nonhuman creatures within this landscape were "nature slaves" who "should be under the rule of a master." Nature, according to Aristotle, "has made all animals for the sake of man."[1]

Aristotle's hierarchical understanding of the subordination of animals to humans was reproduced and "Christianized" in the writings of Saint Thomas Aquinas. Writing in the thirteenth century, Aquinas states the following:

The very condition of the rational creature, in that it has dominion over its actions, requires that the care of providence should be bestowed on it for its own sake; whereas the dominion of other things that have no dominion over their actions shows that they are cared for, not for their own sake, but as being directed to other things ... by divine providence they are intended for man's use in the natural order. Hence it is not wrong for man to make use of them, either by killing or in any way whatever.[2]

While Christian natural law theory could have provided a robust foundation for ethical holism, drawing upon Hebraic notions of stewardship, it became a powerful vehicle for the anthropocentric fallacy. Instead of focusing on the marvelous diversity and wonder of God's creation, natural law theory assumed the ideological trappings of medieval culture, preferring to see the world through the lenses of hierarchy and homogeneity. God's law was etched into the tablets of nature, but human rationality was required to discern it. Hence the apprehension of God's presence in the natural world first had to pass muster before the court of human rationality. This preoccupation with human rationality, combined with certain theological commitments relating to the priority of human beings in the created order, had the effect of sacralizing the anthropocentric fallacy. When natural law theory was slowly stripped of its sacred garb during the Renaissance, the Protestant Reformation, and the Enlightenment, the anthropocentric fallacy survived unscathed. In fact, with the exaltation of human rationality over religious tradition, the anthro-

pocentric fallacy became amplified within progressively secular understandings of natural law and the theories of morality born from the
Enlightenment.

THE ANTHROPOCENTRIC FALLACY IN HOBBES, LOCKE, AND ROUSSEAU

The historical evolution of contractarian thought provides especially
strong evidence of the amplification of the anthropocentric fallacy against
the backdrop of natural law theory. Beginning with the transmutation of
natural law into the concept of ''natural right'' in the thought of Thomas
Hobbes, moral theories based on a social contract or consensus have
understood moral obligation in terms of human interests. Given certain
assumptions about the ''state of nature,'' humans, by virtue of their rational faculties, not only are required to pursue their own interests in the
natural world but also have no need to consult with other co-travelers
who may be affected by human decisions—including God and other life
forms.

Thomas Hobbes' state of nature was a wild and woolly place where
self-preservation could be respected only by the establishment of an all-
powerful earthly sovereign. Once the people enthroned the sovereign
(and thereby surrendered their natural rights to him), the pronouncements
of the sovereign were authoritative and final. The constraints of natural
law were no longer a hindrance to the will and purposes of the sovereign,
as his will was now law. Only the right of self-defense remained—a
right that could not be surrendered by social compact. The Hobbesian
sovereign, who could rule as he pleased, became the model lawgiver—
an exemplary figure of humanity empowered against itself, who answers
to no one and acts according to his interests.

John Locke understood the state of nature in considerably more benevolent terms. Moreover, unlike Hobbes, he entertained an operative
notion of natural law that guided human decisions. Still, Locke embraced
human self-interest as the *locus classicus* of moral reflection. Natural
rights should not be transferred to an all-powerful sovereign; instead,
sovereignty should reside with the people. For Locke, the movement
from the state of nature to civil society requires the institution of private
property—established and preserved by a representative political system.
His concept of private property viewed nature as a vast wilderness without value until humans came along to ''improve'' it with their labor. As
a reward for ''mixing'' their labor with nature, humans earned the right

to own property and, thereby, preserve their individual interests. Hence nature definitionally had no autonomous interests; the terms of the social compact are crafted so that only humans can serve their interests.

The more elaborate notion of the "social contract" by Jean-Jacques Rousseau, and the democratic ideologies it spawned, identified moral and political obligation with the "general will" of the people. We know the right by what the people chose. Hence a social choice process becomes the modern version of the ascent to Mount Sinai. We do not need to search for truth within either tablets of stone or tablets of nature; we need only to determine the will of the people. A social choice process bestows the sceptor of political legitimacy, and the common good of the whole is augered by aggregrating the interests of individuals.

THE ANTHROPOCENTRIC FALLACY IN JOHN RAWLS' *A THEORY OF JUSTICE*

In our time, contractarian thought has reached its apex in the work of John Rawls (1971). The centerpiece of Rawls' theory of justice is the "original position"—the analogue of the state of nature in traditional contract theory. Rawls' original position is a hypothetical construct that is designed to simulate only those conditions that are relevant from the standpoint of justice.

The mythical participants of the original position are charged with the ominous duty of constructing the society in which they will live. Each of the participants has no special relation to the others; they are "mutually disinterested." However, all of the participants labor under a special condition, which Rawls called "the veil of ignorance." The decision makers in the original position have a general grasp of relevant knowledge from political science, economic theory, anthropology, sociology, and psychology. However, the particulars of their society (e.g., its culture, level of economic development, natural resource endowment) and the specifics of their place within the social structure (e.g., their sex, social status, income, educational level) are denied to them. Furthermore, the veil of ignorance denies the participants of the original position any knowledge of their personal traits (e.g., talents, native intelligence, physical strength, the propensity to accept risk).

Given these conditions of impartiality imposed by the veil of ignorance, Rawls argues that the principles adopted by the participants will be morally correct. More specifically, the deliberations of the original position will yield two principles of justice:

First Principle

Each person is to have an equal right to the most extensive basic liberty compatible with a similar liberty for others.

Second Principle

Social and economic inequalities are to be arranged so that they are both (a) reasonably expected to be to everyone's advantage, and (b) attached to positions and offices open to all. (Rawls, 1971: 60)

Although Rawls' theory of justice addresses some of the notable weaknesses of the contract theories of Locke and Rousseau and rescues the concept of the state of nature from anthropological naiveté, it reproduces the anthropocentric fallacy and gives it new respectability. The two principles of justice are principles designed to serve the human community. The veil of ignorance creates an artificial environment of impartiality within a snapshot of biological time; it cannot assure impartiality across time without significant leaps of human imagination (e.g., envisioning what life in the twenty-second century will be like if biodiversity is sharply reduced).

The short-term time preference of the original position is not easily remedied. Rawls recommends that a concern for the welfare of future generations could be achieved within his theory by understanding the parties to the original position "as representing family lines" that are held together by "ties of sentiment between successive generations" (1971: 292). However, this approach suffers from two problems.

First, a generalized sense of responsibility to act on behalf of unborn future generations will likely establish only a weak moral responsibility over a brief moment of biological time, particularly if this is understood to be a responsibility to grandchildren and great-grandchildren and great-great-grandchildren, as Rawls suggests. For example, with an average life expectancy of 75 years and an average age of 25 years for child bearing, the collective lifetimes of four generations span only 150 years.

Second, it is not clear why risk preferences should figure into ecological decisions in the same way that they impact the decisions of the original position with respect to economic institutions and the like, particularly when future generations lack the opportunity to critique the risk preferences of earlier generations. Accepting the risk that one's progeny may be born into the poorest income class of society is considerably different than accepting risks that their habitats may be irreversibly degraded. This is a particularly important point in light of the fact that game theory would suggest that persons within the original position would adopt a more risky attitude toward the future than is suggested by Rawls' risk-averse, maximin solution (Arrow, 1973). More important, there is strong empirical evidence that humans have great difficulty in making integrated probability judgments that adjust risk assessments according to the gravity of potential consequences. More will be said about this momentarily.

Like his intellectual forebears of the contractarian tradition, Rawls commits the anthropocentric fallacy by creating an ethical paradigm that respects the rights and prerogatives of human beings yet disregards the value of other life forms, except as they have value to humans. In the original position, the humpback whale, the chimpanzee, or the elephant have no place around the table. They have no influence over the proceedings; their interests count only if humans decide to represent them.[3]

ANTHROPOCENTRISM IN KANTIAN AND UTILITARIAN ETHICS AND LINGUISTIC ANALYSIS

Similar evidence of the anthropocentric fallacy could be drawn from Kantian ethics or the utilitarian tradition. As suggested in the last chapter, rationality is such a central feature of Kant's theory that it necessarily restricts the sphere of morality to human beings. For Kant, the personhood and dignity of human beings as ends-in-themselves were integrally associated with their capacity for rationality. Animals are ''merely a means to an end.''[4] Humans should not mistreat animals because such abuse may adversely affect the moral character of humans, not because animals have any morally significant interests or rights.[5] More will be said about Kant's theory in Chapter 6 in connection with the theme of human dignity.

While utilitarian theory, as originally set forth by Jeremy Bentham and John Stuart Mill, created a conceptual opening for the moral standing of sentient creatures (because good and bad were defined in terms of pleasure and pain), the demise of cardinal utility at the hands of Lionel Robbins (1932) denied utilitarian theory the opportunity to make any meaningful contribution in this area, particularly in connection with economic theory. (This will be discussed further in the appendix to Chapter 5.) Significantly, even more expansive understandings of utilitarian theory in relation to the animal rights movement failed to develop a moral rationale for ethical obligations beyond sentient creatures (e.g., Singer, 1975).

Even the philosophical approaches to moral discourse that emphasize the relativity of moral concepts and highlight the need for ethical humility fall prey to the anthropocentric fallacy. The legacy of linguistic analysis in philosophy, building on the seminal work of Ludwig Josef Johan Wittgenstein, has understood moral concepts in the context of language games. Although this analysis proved helpful in clarifying the multiple meanings and uses of the language of morality, it also tacitly assumed that morality is confined to language. Only those species who master discursive language (e.g., humans) qualify as either the subjects *of* or the subjects *for* moral discourse. This assumption not only seems unduly restrictive but also ignores the common territory that humans share with other life forms. For example, what about the universal language of DNA—the most remarkable and complex language discovered to date? Why isn't this language relevant for moral discourse? Simply because the genetic code is beyond the scope of normal linguistic analysis to probe, is that any reason to build an anthropocentric bias into linguistic analysis? Indeed, the language of DNA points to a "community of discourse" that is far more nuanced and pervasive than that created by human language (see Dawkins, 1987: 270).

THE MYTH OF DISINTERESTED RATIONALITY

A theme that runs through this chapter is that human rationality evidences a strong bias in favor of human goals and prospects and, accordingly, is not "disinterested" in the conventional sense of the term. Presumably this bias helps to explain why the anthropocentric fallacy is so strongly embedded in the history of ethical theory. Indeed, the history of ethics bears witness to an often inspiring confidence in the power of human rationality to lay bare the main outlines of moral obligation. Unfortunately, though, this confidence has obscured the ways in which human rationality gravitates toward certain perspectives and has succumbed to characteristic sources of distortion.

Philosophers often speak of rationality as a disinterested guide in the pursuit of truth. Perhaps this faith in rationality more generally reflects our enthusiasm for theoretic culture, much like the modern faith in technology. Yet it seems clear that at least three sources of distortion have significantly diminished the degree to which rationality is truly disinterested.

The Problem of Biological Time

The first source of distortion is the tendency to understand biological time in terms of human time (see Wilson, 1984: 120ff). We think of life in terms of the modest time span of a few human generations, whereas the time frame of biospheric life is measured in terms of aeons and millenia. While there is presumably no constraint residing within human rationality that prohibits us from embracing the expanse of biological time, this becomes an impractical place for most of us to live. Moreover, the sheer magnitude of biological time dwarfs our confidence in the worth of human achievement and appears to condemn our lives to insignificance. However, as I will suggest in the final chapter, the constructive embrace of this insignificance is liberating both for human individuals and our species as a whole.

The Problem of Relatedness

The second source of distortion is the tendency to map our relationship to the world in terms of the quality of familiarity (e.g., sensing affiliation with the known and fear with the unknown; interpreting the unfamiliar in terms of the familiar). This tendency is deeply ingrained within human habits of thinking, and its influence is likely never to be fully overcome. At birth the human infant is forced to leave the warmth and security of the womb to experience a strange and glaring world. The transition is traumatic. Everything in this world is unfamiliar. The human infant gradually overcomes its apprehension of the world by progressively recognizing it as familiar, beginning with the warmth of a mother's embrace (see Lumsden and Wilson, 1981: 83–84). From early childhood into adulthood, this mode of learning is mastered and refined, manifesting itself in nearly every aspect of human behavior (e.g., xenophobia, intellectual routinization, stereotyping behavior, the fear of the unfamiliar, a strong preference for "home").

One area in which this habit of thinking is most pronounced concerns how we define communities or relatedness. Humans have progressively understood their ethical relations with other life forms in terms of what Singer (1981) calls the expanding circle, beginning with the family and clan and extending to the village, tribe, the nation-state, and the global human community. One characteristic of this expanding circle of relatedness is the way in which the perception of similarity both precipitates and reinforces the expanding trajectory of moral obligation. This dynamic seems to be evident in the long and arduous struggle for racial and sexual equality among humans. It is also evidenced in the way in which the perception of common interests helps to facilitate the resolution of conflict between large groups. Furthermore, studies from developmental psychology indicate that children typically experience a quantum leap in their capacity for prosocial behavior between ages 4 and 13—a period that coincides with an enhanced ability to empathize with others and to draw connections between their own experiences and the experiences of others (Mussen and Eisenberg-Berg, 1977: 20–23, 66).

Presumably the further enlargement of the expanding circle of moral obligation to encompass the biosphere will depend more on our common *human* interest to preserve the integrity of our biological home than on our ability to find community or to perceive connectedness with non-human life forms. The biological distance between ourselves and an endangered plant or insect species in Amazonia is far too great to be bridged easily by a progressive awareness of similarity. Hence, the traditional association between the perception of familiarity/similarity and one's definition of moral community will not get us very far in terms of biospheric ethics.

In this regard, it is interesting that the pathbreaking developments in animal rights strongly relied on an empathy-based understanding of moral obligation. The animal rights literature in ethics has opened up new avenues for ethical reflection. The leading proponents of this literature argue that animals are deserving of moral regard because of their similarities to humans. Peter Singer (1975) eloquently argued the case for all sentient creatures from the standpoint of utilitarian ethics, and Tom Regan (1983) extended the language of individual rights to mammals alone. Their arguments, although powerful and prophetic, are still laced with anthropocentrism and present a vision of ecological individualism instead of ecological interdependence.

For example, Singer (1979) admits that his utilitarian theory of animal liberation provides no philosophical ground for the preservation of species qua species, and Regan advances the claim that species do not have rights; only individual animals do.

That an individual animal is among the last remaining members of a species confers no further right on that animal, and its rights not to be harmed must be weighed equitably with the rights of any others who have this right. If, in a prevention situation, we had to choose between saving the last two members of an endangered species or saving another individual who belonged to a species that was plentiful but whose death would be a greater prima facie harm to that individual than the harm that death would be to the two, then the rights view requires that we save that individual. (1983: 359)

If the theories propounded by Regan and Singer are correct, the extinction of most plant, insect, and small animal species is of little consequence, because they are too dissimilar from humans. There would be no value in saving an endangered species (e.g., the northern right whale) beyond respecting the rights of the individual whales themselves. This perspective is not only guilty of individualism writ large in the ecological sphere, but also would cheerfully sanction the massive loss of biodiversity across the planet (Callicott, 1989: 39–48; Gunn, 1984). In this case, the perception of similarity in defining moral obligation became the overriding influence and yielded results that can only be viewed as abitrary and capricious from a biological standpoint.

Unfortunately, the perception of similarity also appears to be the operative criterion that guides the investments of the U.S. Fish and Wildlife Services in applying the Endangered Species Act of 1973. For example, between 1989 and 1991, the total spending for species preservation of the bald eagle was $31,330,000; for the West Indian manatee, $17,300,000. By contrast, the investment of the Texas blind salamander or the Choctawahatchee beach mouse amounted to less than $10,000 each—even though they are much closer to extinction (Barro, 1994).

The Problem of Integrated Probability Judgments

A third source of distortion that biases human rationality is that humans apparently lack the intuitive capability to make integrated probability judgments. Humans characteristically confuse low-probability/low-consequence events with low-probability/high-consequence events. In fact, Charles Lumsden and Edward Wilson think that the inability of humans to make intuitive, integrated probability judgments is so pervasive that it may actually have some genetic basis (1981: 88). As a result, humans tend to underestimate the gravity of potential disasters, often not deploying protective measures until a disaster actually takes place. Obviously, this has ominous implications when it comes to ecological issues.

It is instructive, for example, how many policy makers want conclusive proof of global warming before they take serious initiatives in addressing these issues at a policy level. Yet, in light of the potential gravity of global warming, a fully rational response to these issues would not ask, "How can we be sure that CO_2 pollution is creating a greenhouse effect?" Instead, a rational response to such issues would ask, "How can we afford to be wrong?" If there were only a 5 percent chance that CO_2 pollution would precipitate significant changes in the Earth's temperature, it would be a risk far too high to take in light of the consequences of being wrong (e.g., flooding of coastal cities, the death of coral reef systems owing to ocean warming, a loss of biodiversity because of changes in the length of wet and dry seasons, and the inability of species to migrate successfully to cooler latitudes).

These three sources of distortion reinforce that it is appropriate to feel some humility concerning the limitations of human rationality. While we should certainly take some pride in the achievements of theoretic consciousness since the invention of writing, it is equally clear that we cannot credibly conceive of a "disinterested" rationality that functions as an unerring guide for our ethical obligations to planetary life. The myth of disinterested rationality has not served ethical theory well. It has blinded us to critical moral landscapes, obscured our relatedness to the world around us, and spawned a self-destructive and virulent malignancy across the planet.

THE IS/OUGHT GAP AND THE NATURALISTIC FALLACY

Contemporary ethical theory labors not only under the long-standing legacy of anthropocentrism but also under the burden of discredited ethical naturalism. Ethical naturalism refers to ethical theories that use "facts" about the natural world as a foundation for moral judgments. The antinaturalistic bias of the twentieth century—reflected in the ethical philosophies of intuitionism, emotivism, and existentialism—is both a reaction to facile forms of naturalism in the history of ethics and, more immediately, a reflection of the intellectual influence of David Hume (1711–1776) and G. E. Moore (1873–1958).

In the concluding paragraph of section I, part I of book III of his *A Treatise on Human Nature*, Hume argued that moral philosophers have characteristically confused and conflated the realm of fact with that of moral obligation, mistakenly trying to derive an "ought" from an "is."

In every system of morality which I have hitherto met with, I have always remarked that the author proceeds for some time in the ordinary way of reasoning, and establishes the being of a god, or makes observations concerning human affairs; when of a sudden I am surprised to find that instead of the usual copulations of propositions, *is* and *is not*, I meet with no proposition that is not connected with an *ought* or an *ought not*. This change is imperceptible, but is, however, of the last consequence. For as this *ought* or *ought not* expresses some new relation or affirmation, it is necessary that it should be observed and explained; and at the same time that a reason should be given for what seems altogether inconceivable, how this new relation can be a deduction from others which are entirely different from it. But as authors do not commonly use this precaution, I shall presume to recommend it to the readers; and am persuaded, that this small attention would subvert all the vulgar systems of morality, and let us see that the distinction of vice and virtue is not founded merely on the relations of objects, nor is perceived by reason. (1948: 43)

The force of Hume's critique is that the nature of good and evil or right and wrong cannot be discerned from objective facts; such judgments are wholly determined by human sentiments. Hume's celebrated "is/ought gap" obviously poses a critical challenge to all varieties of ethical naturalism.

The "naturalistic fallacy," introduced by G. E. Moore in the first chapter of his *Principia Ethica* (1903), is often associated with Hume's is/ought gap; however, the two should be regarded as distinct issues (Frankena, 1949). Moore contended that moral philosophers have often committed the error of using natural goods (e.g., pleasure) to define a non-natural and unanalyzable property like "good." Because we can no more define or analyze "good" than define the color yellow (we only recognize the property of yellowness when we see it), Moore argued that past philosophers had committed what he termed the "naturalistic fallacy." At the heart of the fallacy is the attempt to define or analyze the concept of "good"—period.[6] However, it became the naturalistic fallacy because moral philosophers often proceeded with this definition by referring to some natural good. One way of identifying the fallacy is that once we define good in terms of a natural quality (e.g., pleasure, self-realization), it is always possible to ask intelligibly whether that natural quality is itself good.

The naturalistic fallacy has achieved a kind of canonical status in contemporary ethical theory. Any hint of ethical naturalism within an ethical theory is sufficient to inspire a charge that Moore's naturalistic fallacy has been committed. Unfortunately, Moore's theory has been interpreted more as an indictment of naturalism than as a cautionary note against the use of the term "good."

One way to respond to the challenge posed by Moore's naturalistic fallacy is to concede his point that "good" is a non-natural and unanalyzable property but working from there to reconstruct a notion of intrinsic good or worth based on the concept of beauty. After all, the primary issue confronting ethical theory at this point is not whether the word "good" is a meaningful term for ethical discourse, but whether there is anything in this world that has intrinsic worth. If there is such an intrinsic good(s), it would provide a foundation for crafting a teleological (or end-based) theory of moral obligation.

Interestingly, the concluding chapter of Moore's *Principia Ethica* hints at such a framework for the concept of intrinsic worth. He asserts that the consciousness of beauty is intrinsically good and lies at the foundation of moral philosophy. Moore writes that

the most valuable things, which we know or can imagine, are certain states of consciousness, which may be roughly described as the pleasure of human intercourse and the enjoyment of beautiful objects. No one, probably, who has asked himself the question, has ever doubted that personal affection and the appreciation of what is beautiful in Art or Nature, are good in themselves; nor, if we consider strictly what things are worth having *purely for their own sakes*, does it appear that any one will think that anything else has *nearly* so great a value as the things which are included under these two heads. I have myself urged . . . that the mere existence of what is beautiful does appear to have *some* intrinsic value; but I regard it as indubitable . . . that such mere existence of what is beautiful has value, so small as to be negligible, in comparison with that which attaches to the *consciousness* of beauty. This simple truth may, indeed, be said to be universally recognized. What has *not* been recognised is that it is the ultimate and fundamental truth of Moral Philosophy. That it is only for the sake of these things—in order that as much of them as possible may at some time exist— that any one can be justified in performing any public or private duty; that they are the *raison d'etre* of virtue; that it is they—these

complex wholes *themselves*, and not any constituent or characteristic of them—that form the rational ultimate end of human action
and the sole criterion of social progress. (1903: 188–89)

Moore's concluding chapter of the *Principia Ethica* is a testament to
holistic thinking in that he brings together the traditionally disparate subjects of aesthetics and morality. The apprehension of beauty or, more
important, the *appreciation* of beauty lies at the foundations of moral
reflection. This conviction stands in direct opposition to a well-
entrenched tradition that ethics and aesthetics are weakly related, if related at all. Take, for example, a passage from Kant's *Metaphysical
Principles of Virtue*:

A propensity to the bare destruction (*spiritus destructionis*) of beautiful though lifeless things in nature is contrary to man's duty to
himself. For such a propensity weakens or destroys that feeling in
man which is indeed not of itself already moral, but which still
does much to promote a state of sensibility favorable to morals, or
at least to prepare for such a state—namely, pleasure in loving
something without any intention of using it, e.g., finding a disinterested delight in beautiful crystallizations or in the indescribable
beauty of the plant kingdom. (1797: pt. 1, par. 17)

Significantly, the relationship between the definition of moral obligation and the apprehension of beauty is one of the emergent insights of
the environmental ethics literature. Aldo Leopold's pathbreaking *Sand
County Almanac* (1949) is a monument to an organic grasp of nature's
beauty. More recently, Holmes Rolston's ethical explorations of the concepts of "wildness" and "wilderness" have unearthed the rich moral
and aesthetic symmetry of natural beauty (1986: 118–42, 180–205). A
similar theme is echoed in Peter Miller's (1982) concept of "value as
richness." Likewise, Richard Austin (1985), drawing on the thought of
Jonathan Edwards, has articulated an engaging framework for thinking
about ethics and natural beauty, and Robert Elliot (1989) has employed
the metaphor of vandalism in describing the wrongfulness of environmental degradation in relation to aesthetic considerations.

Building upon the close association between the moral and aesthetic
realms of life charted by Moore, the classic is/ought gap identified by
Hume could be significantly narrowed by appealing to Hume's own theory of the moral sentiments. As J. Baird Callicott persuasively argues,

if human moral sentiments "are both *natural* and *universally distributed* among human beings," then Hume's moral theory does not "collapse into an emotive relativism."(1982: 167–68), The claim, for example, that we as human beings have a moral obligation to respect the essential interdependence of the biosphere could be defended on Humean grounds by appealing to the strong "preference for life" that seems to be genetically "hardwired" within human sentiments. Therefore, Hume's divide from is to ought could be crossed, in a circuitous manner, by appealing to the widespread preference to life exhibited among humans. This argument receives exceptionally strong support in the concept of "biophilia," advanced by biologist E. O. Wilson. We will now explore the notion of an innate human preference for life and then return to explore an alternative bridge across the chasm between is and ought.

BIOPHILIA

In an engaging and lucid book called *Biophilia* (1984), Harvard biologist Edward O. Wilson explores the innate attraction of humans to life. The central premise of the book is that humans will place greater value on themselves and on other life forms the more that they understand the immense wonder and diversity of the biological life of Earth.

Wilson contends that because the bounty of biological life that surrounds us was here before us, we have taken for granted the remarkable beauty and diversity of the biosphere. As newcomers in Earth's evolutionary history, we evolved within the diversity of life and "have never fathomed its limits."

As a consequence, the living world is the natural domain of the most restless and paradoxical part of the human spirit. Our sense of wonder grows exponentially: the greater the knowledge, the deeper the mystery and the more we seek knowledge to create new mystery. This catalytic reaction, seemingly an inborn human trait, draws us perpetually forward in a search for new places and new life. (1984: 10)

Yet, as Wilson notes, our impulse to master nature and our fascination with our own capabilities to manipulate nature have obscured the wonder and glory of all that was here before our arrival. For example, the labyrinthine life that lies under our feet is a marvelous biological landscape that we irreverently dismiss as "dirt."

Think of scooping up a handful of soil and leaf litter and spreading it out on a white ground cloth, in the manner of a field biologist, for close examination. This unprepossessing lump contains more order and richness of structure, and particularity of history, than the entire surfaces of all other (lifeless) planets. It is a miniature wilderness that can take almost forever to explore. . . . Our lump of earth contains information that would just about fill all fifteen editions of the *Encyclopedia Britannica*. (13–14, 16)

Instead of savoring the "cathedral feeling" of rain forests, we have elected to burn "a Renaissance painting to cook dinner" (25).

Wilson suggests that the *biophilia* evidenced by humans (i.e., their preference for life) may have a genetic basis in our evolutionary journey from the forest to the savanna. The expanse of the savanna offered many new opportunities for our ancestors.

First, the savanna by itself, with nothing more added, offered an abundance of animal and plant food to which the omnivorous hominids were well adapted, as well as the clear view needed to detect animals and rival bands at long distances. Second, some topographical relief was desirable. Cliffs, hillocks, and ridges were vantage points from which to make a still more distant surveillance, while their overhangs and caves served as natural shelters at night. During longer marches, the scattered clumps of trees provided auxiliary retreats sheltering bodies of drinking water. Finally, lakes and rivers offered fish, mollusks, and new kinds of edible plants. Because few natural enemies of man can cross deep water, the shorelines became natural perimeters of defense. (110)

For all organisms, the choice of habitat is a critical decision. "If you get to the right place, everything else is likely to be easier" (106). If other animals are guided by "inborn rules of behavior" that help them select the proper habitat, it is likely that the human preference for life is a product of natural selection that gave our ancestors an evolutionary edge in selecting savanna habitats most conducive to their existence. If so, this would likely be evidenced in the aesthetic judgments of humans (see also Eibl-Eibesfeldt, 1989: 613–14, 674). Beauty is a window into our biological history.

Among humans, the biological basis of beauty is most pronounced in the way in which people select their habitats or artificially replicate savanna-like features in their living spaces. Wilson notes that typically

> whenever people are given a free choice, they move to open tree-studded land on prominences overlooking water. This worldwide tendency is no longer dictated by the hard necessities of hunger-gatherer life. It has become largely aesthetic, a spur to art and landscaping. Those who exercise the greatest degree of free choice, the rich and powerful, congregate on high land above lakes and rivers and along ocean bluffs. On such sites they build palaces, villas, temples, and corporate retreats. Psychologists have noticed that people entering unfamiliar places tend to move towards towers and other large objects breaking the skyline. Given leisure time, they stroll along shores and river banks. . . . When people are confined to crowded cities and featureless land, they go to considerable lengths to recreate an intermediate terrain, something that can tentatively be called the savanna gestalt. At Pompeii the Romans built gardens next to almost every inn, restaurant, and private residence, most possessing the same basic elements: artfully spaced trees and shrubs, beds of herbs and flowers, pools and fountains, and domestic statuary. When the courtyards were too small to hold much of a garden, their owners painted attractive pictures of plants and animals on the enclosure walls—in open geometric assemblages. (110–11)

The same dynamic is strikingly evident in the legacy of Persian rugs, in that traditional Bedouin peoples understood their colorful and ornate rugs to be traveling gardens that could be taken into the desert.

Human *biophilia* is so strong within the human breast that even the most perfect recreations of our environment would be "empty" if they were voided of life. The most intricate holographic representation of a life-full environment—such as the "holodeck" projected by Star Trek's visionaries—can only mimic the satisfaction we find from life; it cannot be a substitute. If the most elaborate life support system could reproduce a fully perfect yet fully artificial earth habitat, the humans who occupied this environment would surely understand their world as fully dead—merely a tomb with creature comforts. Life without life would surely be infinitely more painful than life without purpose.

Wilson's concept of *biophilia* gives substance to Callicott's (1982) claim that Hume's celebrated is/ought gap, when understood from the standpoint of his theory of moral sentiments, can be crossed. That which is is also what ought to be, because we are too deeply rooted in the biological life of the planet to erect artificial distinctions between "homeness" and "goodness." Moreover, the notion of *biophilia* is profoundly destructive of attempts to draw firm boundaries between the realm of aethestics and morality. Boundaries that are sharply delineated are also likely to be arbitrary boundaries.

Furthermore, while the essential truth of Moore's famous naturalistic fallacy remains unchallenged concerning the undefinability of the concept "good," the concept of *biophilia* similarly constrains us to agree with the neglected insights of *Principia Ethica*'s final chapter: The impossibility of defining good does not obstruct our quest to determine that which is inherently good. We have grown up in a complex and wondrous matrix of life. We can no more deny the goodness of the life that surrounds us than deny the worth of the most celebrated achievements of the human imagination.

Yet, having said this, we should appropriately be apprehensive about touting naturalistic "solutions" to Hume's is/ought gap. Is it not always possible to ask, with Moore, whether *biophilia*, as a presumably genetic trait, is good? What guarantee is there that a human sentiment as pervasive and beneficial as this one is a morally "correct" sentiment? If, for example, humans, by some fluke of evolution, were genetically hardwired to prefer lifeless habitats and ecological monotony, would that make it right? Obviously not! What place, then, does an argument based on human sentiments, no matter how well distributed, have in environmental ethics? Is this not simply a much more subtle restatement of the anthropological fallacy? Is there not anything that is intrinsically good in the universe that does not first depend upon the approval of our moral, aesthetic, or theological sensibilities?

THE PRIME DIRECTIVE

The arbitrary boundary between morality and aesthetics obscures a more fundamental Berlin Wall that philosophers have erected between creativity and morality. Creativity has always occupied a cherished place in the human aesthetic consciousness. To my knowledge, though, it has never been embraced as an essential attribute of morality, except as it has been associated with the attributes of God in theological ethics and theoretical constructs in process philosophy.[7] Unlike the aesthetic dimension of life, the value of creativity stands by itself; it does not need to be mediated through human consciousness. It is possible to speak cogently about nature's creativity in a way that we could not refer to nature's aesthetic judgment—particularly when creativity is defined more restrictively in terms of generativity.

The intrinsic goodness of life does not spring from the altruistic behaviors of its members or nature's benevolent aura, but instead from its boundless generativity. The natural world presents both kindness and cruelty, but it flaunts possibility. While life's creativity operates within bounded ecological limits, the possibilities of life seem seductively boundless, feeding both the illusion of nature's indestructability and the dream of transplanting life throughout the universe.

What creature, with even the most modest semblance of consciousness, could fail to marvel at the generative capacities of life? The natural world, as we experience it within Earth's biosphere, teems with mind-numbing intricacy and diversity. Who can fathom its depths? Nature's capacity to regenerate itself in the face of destruction amazes even the most seasoned biologists. The rapid return of flora and fauna after the eruption of Mount Saint Helens is a case in point. A much more remarkable example is the biological recovery of what remained of the island of Krakatau (misnamed Krakatoa) following the world-shaking eruption of 1883. Today a thick green forest, populated by a remarkable array of insects and animals, covers the once lifeless island (Wilson, 1992: 16–23).

The remarkable processes of speciation and natural selection have demonstrated creativity that boggles the human imagination. Who can fail to experience awe by the wonder of nature? The replication of DNA, the sonar systems of bats and dolphins, the evolution of the eye, electric fish—all attest to the spectacular creativity of life (see Dawkins, 1987). No ecological niche has been left untouched; no possibility, it seems, has been unexplored.

When the Berlin Wall between creativity and morality is dismantled, it seems quite ridiculous to force the claim that everything must be processed through human consciousness before it takes on value. The generativity of the natural world is an intrinsic good that no intelligent life form could rationally deny without also denying its own existence. We owe our own biological existence to the generative powers of Earth's biosphere. Without the potential unleashed by nature's generativity, the achievements of theoretic consciousness from gene mapping to space exploration would be quite impossible. If there is intelligent life on other planets, we could imagine that they would be forced to the same conclusion.

Hence, the generativity of life—in all its fullness and diversity—has profound moral significance. To respect life is to affirm the worth of life's creativity, and to respect life's creative potential in both its "constructive" and "destructive" dimensions is to affirm our place in the biosphere. This clearly must be the foundational premise for moral reflection and the ethical enterprise. It also forms a durable bridge on which to cross Hume's chasm between is and ought. In this sense, life is the earthly *summum bonum* that orders all things to it. It refers to the totality of life within the biosphere and the complex interactions of Earth's life forms. Hereafter I will capitalize the term "life" to distinguish its use from a consideration of particular life forms (e.g., chimpanzees, dolphins, humans, spruce trees).[8]

The pinnacle of consciousness on our planet is the theoretic consciousness of *Homo sapiens* (see Chapter 2). This too is a reflection of LIFE's creativity. The remarkable growth of the human neocortex and the evolution of theoretic consciousness did not spring from nowhere; it evolved (or was created) in the context of LIFE. As a product of both genes and culture, theoretic consciousness among humans can neither exclude itself from LIFE nor legitimately assert its prerogatives over LIFE. Theoretic consciousness is a resplendent representative of LIFE's possibilities; it too should open up more possibilities, not close them down.

One possibility opened by human consciousness is discursive moral reflection. Drawing on both the resources of mythic consciousness and theoretic consciousness, humans have developed elaborate means to regulate their interactions with one another. Choice creates dilemma, and the awareness of choice among humans has posed seemingly countless moral dilemmas for humans to solve by means of religion, law, and philosophical discourse. The significance of this cultural development can hardly be overstated.

Whereas moral reflection, at the level of theoretic consciousness, took a decidedly anthropocentric turn in the history of Western ethics, it did not have to be so. If, for example, Aristotle had been able to consult the conceptual resources of contemporary microphysics, macrophysics, evolutionary biology, and neurobiology, his grasp of moral decision making would be quite different. Would he have ordered all things according to human rationality, regarding human prerogatives and purposes as the crown of the natural world, or would he have located human rationality within the broader landscape of LIFE, embracing its uniqueness without denying its evolutionary continuity to LIFE?

A moral theory that partakes of the anthropocentric fallacy is a moral theory that closes down possibility. Certainly the most vivid example of this is the disastrous decline in biodiversity across the planet. With extinction rates as high as 27,000 species per year in rain forests *alone* (Wilson, 1992), it is as if a cataclysmic plague has brought darkness to the Earth. This time, though, the Dark Age is the result of the cancerous colonization of human cultures, not the bubonic plague. The malignancy that invites Dark Ages and closes down possibilities also, paradoxically, enjoys the blessing of theoretic culture. That theoretic consciousness— a product of LIFE's possibilities that epitomizes LIFE's possibility— could be seduced to sanction the closing down of possibilities is an irony almost too painful to grasp. While the "myth of objective representation" (see Chapter 2) helps account for the casual relationship between theoretic consciousness and anthropocentrism, once the myth is unveiled and seen for what it is, the relationship seems forced and "unnatural" in the fullest sense of the term.

The respect or reverence for LIFE is central to our most fundamental moral intuitions; the denial of LIFE is most basic to our conceptions of evil. The respect for LIFE is also the emergent ethos of the most sophisticated strains of theoretic culture—from quantum mechanics *to* evolutionary biology *to* astrophysics. One cannot become aware of the profound interconnectedness of life without also affirming the value of LIFE. This portrait of connectedness takes on distinct moral hues where the evolution of ratiocination permits reflection on the choices open to intelligent life forms. The intrinsic goodness of LIFE is not a mere datum in the moral calculus of a species; it must necessarily be a foundational point of departure for coherent ethical reflection, a place where fact intersects value and the re-presentation of LIFE is also the reaffirmation of LIFE. Without such an ontological ground for ethics, we are lost minds casting about in a vast sea of life, pathetically struggling for control over that which cannot be controlled without also being destroyed.

The respect for LIFE is at one and the same time the respect for the essential interdependence of the biosphere. It is the intrinsic goodness of nature's generativity that grounds the worth of biospheric interdependence—as ecological competition is a prerequisite for biospheric interdependence and biospheric interpendence is an essential precondition for generative potential of our biosphere. Without the fierce competition for scarce resources and the constant interplay of life and death struggles, our biosphere would lose its astounding generativity. Ecological monopoly is antithetical to nature's generativity. It would be just as unseemly for a species of intelligent mosquitoes to overrun the world, destroying habitats and depopulating other species, as it is for human beings.

Albert Schweitzer's famous principle of "reverence for life" is a denial of LIFE in the sense that it is used here. The reverence for life, in Schweitzer's use of the term, focuses on the intrinsic and infinite value of *individual* life forms within the biosphere. By contrast, the reverence for LIFE focuses on the intrinsic value of ecological interdependence. The value of individual life forms within a reverence for LIFE perspective is ambiguous and finite in that no single life form or species should place its own life interests ahead of LIFE itself. As such, neither humans nor animals have rights in the sense that they can legitimately press exclusive and absolute claims for the interests and prerogatives of their species over and against the interests of other species. Extreme versions of both human rights and animal rights deny LIFE its interdependent constitutional charter.

As members of the human species, we exist within a fragile and exceptionally diverse web of life which we call the biosphere. We inhabit the planet and partake of both its abundance and travail, along with countless other plant and animal species. The essential interdependence between our species and these other plant and animal species is an indisputable fact of life. Within an interdependent world, no species has the ecological prerogative to monopolize the Earth's resources or possesses a legitimate veto power to eradicate the genetic heritage of other species. To deny the essential interdependence of Earth's biosphere is to repudiate our own stake in the future of the planet.

Our primary moral obligation to LIFE is not to manage or even to promote it, but to avoid destabilizing and destroying it. The principle "do no harm" has no greater relevance than here. Our experience to date with "managing" LIFE by controlling habitats has been abysmal. To manage the complex interdependence of LIFE is to reduce the diversity of LIFE. This is not only because we are forced to pick winners and losers (coronating some creatures and vilifying others), but also because the complexity of ecosystems is so great that benevolent human intervention is likely to destabilize habitats and thereby reduce diveristy (Leopold, 1949: 133–37; Ehrenfeld, 1991). Moreover, the impulse to manage necessarily seeks conditions that are conducive to control (usually the same conditions that discourage diversity). Harmony, for managers, is typically synonymous with homogenization.

Nowhere is this dynamic more pronounced than in "modern" agriculture. Productive efficiency in modern agriculture requires tremendous energy inputs in the form of mechanical equipment, fertilizers, and pesticides, as well as clearing vast tracts of land (i.e., destroying "undesirable" life), reducing the diversity of land to soil—its least common denominator. This leveling process of modern agriculture stands in stark contrast to that of traditional agriculture, which embraces considerably more ecological diversity and is much more energy efficient than modern agriculture (i.e., it generates an overall positive caloric output in comparison to the strongly negative calorie balance of modern agriculture).

To respect LIFE, then, is to refrain from harming it. The awareness that our moral responsibilities to the environment are largely "negative" in character is both a recognition of the intricate complexity of ecological interdependence (i.e., control yields chaos) and an affirmation that the possibilities created by LIFE are far richer than the possibilities that we create for ourselves by capping LIFE's potentiality. Our attempts to control LIFE have a curious way of defeating LIFE. It would be better to stand aside and enjoy the music of LIFE instead of feeling compelled to pick up the baton and to conduct the orchestra.

Our present posture toward LIFE is best described by the metaphor of cancerous colonization. The biosphere is our victim; we suck life from it indiscriminately by overwhelming it. Our biological impact is far beyond any predatory need of our species for sustenance. Our modus operandi is that of marauding ecological looters who confiscate resources, destroy habitats, and unload pathogenic wastes across nearly every ecological zone. Like cancer cells that insidiously destroy healthy cells, our conception of life destroys the diversity of LIFE. Where we arrive, life dies.

The Prime Directive of LIFE is to affirm the generativity, diversity, and interdependence of LIFE—the goodness of LIFE. Simply stated, the Prime Directive is: *Respect the essential interdependence of the biosphere.* To comply with the Prime Directive of LIFE is to no longer be agents of earth cancer. Living in accordance with the Prime Directive of LIFE is to rediscover life. It requires us to engage in a wholesale reformation of academic disciplines such as ethics and economics, and to ask fundamental "religious" questions about who we are and about the meaning and purpose of human life.

The ability to celebrate the wonder and diversity of LIFE is the birthright of every human being. For cancerous colonizers, the Earth is something to be devoured—a condemned prisoner. For human beings who celebrate their humanity, the Earth is filled with awe and discovery—a landscape for liberation. To respect the Prime Directive is to participate in "the liberation of life."[9]

The anthropocentric fallacy has taken a serious toll on ethical theory. The legacy of the anthropocentric fallacy in ethics has not only reduced the essential goodness of LIFE to restrictive "good for us" propositions, but, more perversely, has darkened the human imagination by erecting Berlin Walls that have lulled us into a malignant relationship to the biosphere. Enlightened democracies have lapsed into a totalitarianism of human interest. Market economies have become proficient in creative accounting, depleting nature of its "capital" resources and accounting for it as "income." Contrary to those who cheerfully trumpet the "end of history," the excesses sanctioned by the anthropocentric fallacy threaten to end history.

NOTES

1. Aristotle, *Politics*, 1254b and 1256b.

2. Thomas Aquinas, *Summa Contra Gentiles*, trans. by the English Dominican Fathers (New York: Bentigen Brothers, 1928) bk. 3, pt. 2, chap. 112. Quoted in Johnson (1991: 18–19).

3. It should be noted that Brent Singer (1988) has suggested some modifications in Rawls' theory that would strengthen its capacity to support an environmental ethic.

4. Immanuel Kant, *Lectures on Ethics*, trans. by L. Infield (New York: Harper Torchbooks, 1963), pp. 239–40.

5. Immanuel Kant, *Metaphysical Principles of Virtue*, trans. by James Ellington (Indianapolis, IN: Bobbs-Merrill, 1964), pt. 1, par. 17.

6. See Warnock, 1978: 14. Frankena (1949) also argues that it is not to be termed a "fallacy" in the proper sense of the term.

7. For example, the theme of redemption is a profoundly ethical *and* ecological concept in Christianity (e.g., Romans 8:18ff), which is expressly related to the belief in God's continuing participation in creation by acts of re-creation.

8. Birch and Cobb employ a similar device in their discussion of LIFE as an ethical and religious concept (1981: 176–202).

9. The choice of words here highlights my debt to the work of Birch and Cobb (1981), especially Chapter 6, "Faith in Life."

Chapter 5

The Economics of Earth Cancer

THE BERLIN WALL BETWEEN HOUSEHOLD AND HABITAT

Mainstream economic theory cannot see, let alone account for or correct, the pernicious proliferation of earth cancer. Its vision has been gravely obscured by the Berlin Wall it has built between the human "household" and the human "habitat."

The original derivation of the word "economics" in Greek refers to the proper management of the household. Logically, it is nonsensical to speak about the proper management of the household without also addressing the larger context of human habitats. Yet the myopic preoccupation of economics to satisfy human preferences, combined with the virtual marginalization of ecological issues, has forced a separation between household and habitat. One of the primary consequences of this divorce is a double standard concerning the worth of competition. Whereas mainstream economic theory proudly boasts of the value of competition in relationships within the human household, it legitimates and even cheers monopolistic modes of production and allocation when it comes to the ecological sphere.

Since David Ricardo's seminal analysis of how the corn laws in England benefited the interests of aristocratic landowners to the detriment of both energetic capitalists and common laborers in his *Principles of Political Economy and Taxation* (1817), economic thought has taken a decidedly negative outlook on all forms of economic parasitism. According to Ricardo, the wealthy landowners lived off the ingenuity and effort of others—contributing nothing to overall economic productivity and charging rent for the natural productivity of the land. By keeping the price of corn high, the corn laws benefited the interests of the landowners and disadvantaged practically everyone else. Ricardo's dispassionate analysis of the parasitic function of the corn laws later found more passionate expression in neoclassical antipathy toward coercive monopolies that prey on unwitting consumers and Marxist contempt for the exploitation of labor and economic colonialism.[1]

It is curious that the economist's aversion to noncompetitive and coercive economic relations has rarely found expression when it comes to matters of the environment. And yet it is the relationship between humans and the biosphere where coercive and cancerous colonization becomes the most pronounced and detrimental in terms of long-term economic productivity.

One suspects that this monumental oversight ultimately stems from the anthropocentric bias that economics inherited from its ethical heritage—in particular, the legacy of utilitarianism (see the appendix to this chapter). Because the disciplinary vision of economics is preoccupied with households to the exclusion of habitats, economic theory has pronounced its blessing upon the abduction of all the natural features of the world and placed them in the service of (human) "utility." Land, lakes, oceans—indeed entire ecosystems—are labeled "natural resources." Humans are thereby encouraged to accept a totalitarian perspective on the biosphere, denying "non-human" entities their intrinsic value. It would be difficult to conceive of a more thoroughgoing totalitarian ideology than that of contemporary economic theory. The common thread running through the neoclassical, institutionalist and Marxist schools of economic thought is that economics finds its raison d'être in human purposes and preferences.

Ecological monopolies have the same deleterious effect on production and consumption in human habitats as do monopolistic economic relations in human communities. It corrupts the entire meaning of competition to tout the virtues of competition for the human household and embrace a monopolistic state of affairs for the human habitat. Appropriately, the leading insight from the budding discipline of ecological economics is that humans must learn to live within the Earth's ecology, not over it. Yet even within the promising framework of ecological economics, one detects indications of monopolistic thinking. For example, the promising work on environmental accounting, focused on the depreciation of environmental capital resources (see El Serafy, 1988, 1991), still reinforces the traditional monopolistic pattern of thinking, in that the worth of a habitat can never be fully captured by even the most sophisticated system of national accounts. Habitats should not be places where monopolistic interests are permitted to roam free if the price is right; they are vital centers of ecological competition.

The profoundly competitive nature of the biosphere is mirrored in every habitat on our planet. Where humans occupy a habitat, they are one species among many others—often competing directly with other species for the same resources. To assert monopolistic control within the naturally competitive venue of human habitats is to cripple the productivity of the biosphere in the same way that the coercive restrictions and notorious trade barriers of mercantilism stifled the economic productivity of Adam Smith's world. Economic theory remains badly in need of a new *Inquiry into the Nature and Causes of the Wealth of Habitats.*

HABITATS AS MARKETS

Virtually any spot on our planet's surface can be redefined as a complex, multilayered market—what I will refer to as an "ecological market." Creatures of all shapes and sizes place demands on environmental resources in much the same way that human consumers express their preferences in the marketplace.

In both traditional economic markets and habitat-based ecological markets, the interplay of supply and demand determines a market equilibrium. In economic markets, this equilibrium is expressed in terms of the price and quantity of goods and services. In ecological markets, the equilibrium between supply and demand represents the complex interaction among the countless life forms who inevitably take on the role of being both consumers and suppliers throughout their life cycles. Concepts like "the food chain" or "energy cycles" are windows into this complex interaction of supply and demand. On our planet the most complex ecological markets can be found within the canopies of the rain forests and the oceans.

The primary difference, of course, between economic markets and ecological markets is that the quality of voluntariness is missing in ecological markets.[2] Consumers do not "vote with their dollars" in ecological markets but instead express their preferences in terms of the familiar categories of predation, expropriation, and colonization. Similarly, in ecological markets suppliers are not a cohesive set of actors who respond to the demands of a dynamic marketplace. More often than not, suppliers are unwitting (and unwilling) participants in a life and death struggle for survival.

One striking similarity between economic and ecological markets is that both types of markets are dynamic in character, having equilibria that are constantly in flux. However, whereas this characteristic is explicitly recognized in economic markets, it has often been downplayed in ecological markets. The term "balance of nature" in ecology has deceptively static overtones. When the father of modern ecology, Charles Elton, wrote that "the balance of nature does not exist and perhaps never has existed" (1930: 17), he was referring to the tendency of scientists to think of habitats in static—almost Platonic—terms. In reality any habitat is the result of dynamic forces at play; it is misleading to conceive of nature as a delicate vase that should never be taken off the shelf and handled.

Another similarity between economic and ecological markets is the way in which competition promotes the health of both types of markets. Robust economic *and* ecological markets are characterized by both competition and diversity. Just as monopolistic economic markets are characterized by bland product uniformity, monopolistic ecological markets are noted for their lack of biological color.

Most of the world's urban centers are monuments to ecological monopolism. The biological blandness of the world's cities—while being partially offset by parks, bright lights, and the excitement of human interaction—cannot help but take its toll on the human spirit. For those affluent enough to arrange for occasional getaways to the countryside or to recreate some measure of urban biodiversity through gardens, houseplants, and pets, the monotonal character of city life is not oppressive. However, for low-income households, the city has a distorting influence on life—with the cockroach and rat as its biological beneficiaries. Green is engulfed by gray. Even modern agriculture is a testament to biological monotony. The pronounced lack of biodiversity associated with mechanized farming, combined with overreliance on pesticides, has produced a kind of cultivated colorlessness which usually escapes the attention of the casual rural visitor (Leopold, 1949: 185, 188–89).

A habitat-centered approach to economics celebrates the value of competition while being fully sensitive to the inherent environmental constraints of an ecological zone. A habitat-centered approach, unlike contemporary economic theory, does not reduce land to an undifferentiated residual category which functions as a proxy for all finite resources (see Daly and Cobb, 1989: 105ff). Instead it tries to capture the distinctive characteristics of an ecological zone and builds on its "comparative advantage." The "parcel-oriented thinking" of mainstream economics is replaced by "landscape thinking," where the complex inter-relationship between human communities and ecological zones can be fully appreciated.

From a habitat-centered perspective, the familiar tradeoff between economic growth and environmental protection—or jobs versus the environment—is a false dichotomy. The real tradeoff is between ecological monopolism and ecological competition. Monopolistic approaches to economic growth cannot help but yield uneven and inefficient growth paths. It is only when competitive values are integrated within both economic *and* ecological markets that long-term economic growth will have a stable foundation.

MONOPOLISM AND ENVIRONMENTAL TRANSFORMATION

The distorting effects of ecological monopolism are most clearly evident in the way that human communities have gone about transforming their respective habitats. Since the invention of agriculture and the related phenomenon of urbanization nearly 10,000 years ago, humans have devoted considerable energy and creativity to remaking their environments. Most often this environmental transformation has taken the form of clearing land for cultivation and the construction of shelters. More recently, though, the human passion for environmental transformation has taken the form of destabilizing and poisoning entire ecosystems through an extravagant menu of technologies.

The problems of deforestation, global warming, the depletion of the ozone layer, acid rain, the pollution of the world's oceans, and massive species extinctions represent a new generation of crises associated with environmental transformation. Even the ways that modern humans manipulate the environment to secure their physical means of sustenance has a degrading impact on the integrity of our planet's ecology and the well-being of its inhabitants. The excessive use of pesticides and irrigation, combined with the grotesque institution of the factory farm and the macabre ruins left by slash-and-burn agriculture, are typical of the unwholesome impact of humans upon the environment. This is to say nothing of our penchant for destroying "undesirable" or "insignificant" animal and plant species because they do not fit our habitat preferences or for overharvesting economically "profitable" species to the point of extinction.

Environmental transformation has a price tag associated with it. In the thought of John Locke, the cost of transformation was measured in terms of the human labor it required. Moreover, for Locke, the very act of "mixing" one's labor with nature was all that was necessary for transforming natural endowments into "private property." Since Locke's severely anthropocentric attitude toward nature, economic thinking of both neoclassical and Marxist vintage has adopted a "parcel perspective" to the biosphere, reckoning the cost of environmental transformation in terms of the market price of the required inputs for change.

Modern cost-benefit analysis is a case in point. Cost-benefit analysis understands the price of environmental transformation in terms of the mix of utilities and disutilities for human communities and their future generations. Accordingly, the costs and benefits associated with a habitat transformation is measured in terms of its anticipated rate of return or impact on human aesthetic or recreational values. The fact that a habitat is home to many creatures and contributes to a larger ecological landscape is of no great consequence. Moreover, when standard "present value" calculations are factored into a cost-benefit analysis, the net effect is to bias environmental transformation in the direction of short-term financial goals, owing to the fact that the discount rate on capital resources is invariably higher than the "shadow" discount rate for ecological resources. For example, at a long-term interest rate of 10 percent, according to present value calculations, it would not be economical to invest more than $36,283 today in order to avoid an ecological loss of $500 million in 100 years. As such, present value analysis and cost-benefit analysis are exquisite tools for ecological monopolism (see Cairncross, 1992: 30–35).

One way of exploring the effects of ecological monopolism is to think of habitat transformation in terms of the familiar categories of supply and demand (see Figure 5.1). Let us measure the price of habitat transformation (P) on the vertical axis and the degree of transformation (T) on the horizontal axis.

Assuming a consistently inverse relationship between the price of transformation and the rate of transformation, the demand curve among humans for habitat transformation (D_{hum}) can be represented by a downsloping curve. The lower the price associated with habitat transformation, the more transformation will be demanded; the higher the price, the less transformation will be demanded. By contrast, the demand curve for all other species (D_{aos}) presumably would be perfectly inelastic—as price considerations would not be a factor in their collective preferences for habitat transformation—and would be inherently conservative (i.e., the more stable a habitat the better).[3]

The supply curve for habitat transformation (S_{hab}) is a kind of "habitat profile," representing both the financial *and* shadow costs[4] associated with the transformation of a particular habitat. The type of environmental transformation envisioned could be as limited as a housing project, a dam, or a highway or as massive as the destruction of a tropical forest or long-term climate change. The supply curve for habitat transformation is upsloping, like the supply curves of economic markets, reflecting the familiar law of supply. In this case, suppliers will produce more units of environmental transformation at higher prices. However, unlike traditional supply curves, the supply curves for habitat transformation (S_{hab}) incorporate the shadow costs associated with habitat transformation— costs felt by the ultimate "supplier" of habitat transformation (i.e., the environment itself) but not assumed by the individual firms producing the transformation. Therefore one would expect that habitat supply curves (S_{hab}) will become progressively steeper as the degree of transformation increases. This would be primarily due to the "increasing costs to scale" associated with environmental transformation, particularly when shadow costs (e.g., species extinction, pollution of air and water resources) are accounted for. By contrast, traditional supply curves tend to flatten out due to "decreasing costs to scale."

One would imagine that the creation of environmentally friendly technologies would shift the entire supply curve to the right, permitting increased levels of environmental transformation at lower prices. Moreover, it goes without saying that the supply curve for habitat transformation may vary dramatically from one habitat to the next. For example, the supply curve for a tropical rain forest would presumably be very steep, reflecting the immense costs associated with the extinction of species and the permanent destruction of fragile laterite soils. Even though the direct financial costs associated with the destruction of a forest may be relatively insignificant (e.g., slash-and-burn agriculture), the shadow costs—depending on the system of valuation used—could be staggering.

In Figure 5.1, the equilibrium level of transformation for humans is T_1 and the equilibrium price is P_1. This compares with an equilibrium level of transformation of T_2 for all other species (D_{aos}). The distance between T_1 and T_2 represents the extent of monopoly power held by

Figure 5.1
Supply and Demand Aspects of Habitat Transformation

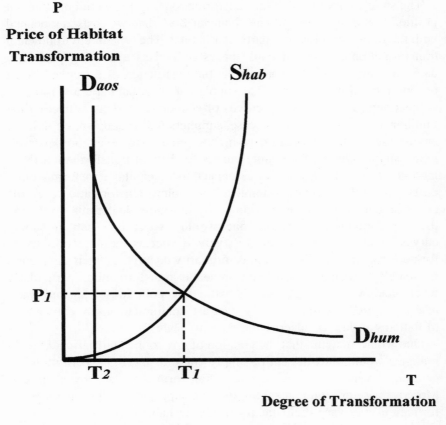

P

Price of Habitat
Transformation

D_{aos} S_{hab}

P_1

D_{hum}

T_2 T_1

T

Degree of Transformation

humans with respect to the degree of habitat transformation. It goes without saying that *some* degree of monopoly power will always be present in habitats managed by humans; the complete absence of monopoly power would presumably require a return to our hunter-gatherer days. However, such power could be exercised in a benign manner, particularly if the shadow costs associated with habitat transformation are accounted for and factored into economic decision making.

The most obvious indicator of monopoly power in the transformation of habitats concerns the shape of the habitat supply curve (S_{hab}). Present conventions in cost-benefit analysis, for example, would rely on direct financial costs associated with habitat transformation, neglecting most, if

Figure 5.2
Habitat Supply Curves with and without Shadow Costs

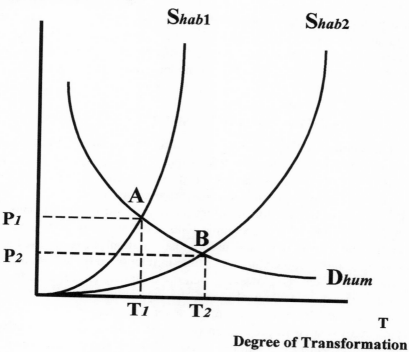

not all, shadow costs, except perhaps those associated with human recreational preferences.

Figure 5.2, for example, depicts two alternative supply curves for habitat transformation and one demand curve for humans. As in Figure 5.1, the price of habitat transformation is measured on the vertical axis and the degree of habitat transformation is measured on the horizontonal axis. The first habitat supply curve (S_{hab1}) describes both the financial *and* shadow costs associated with increasing levels of environmental transformation. It yields an equilibrium level of transformation of T_1 at an equilibrium price of P_1. By contrast, the second habitat supply curve (S_{hab2}) includes only financial costs. It is to the right of the first curve and is considerably flatter, yielding an increased equilibrium level of transformation (T_2) at an artificially lower price (P_2). As a consequence, a higher (and presumably more destructive) degree of environmental transformation is encouraged, at the expense of the habitat and its inhabitants, because of deficient accounting techniques at best, or ecological irresponsibility at worst. In effect, this involves using our monopoly prerogatives to extract subsidies from the environment. In Figure 5.2, this environmental subsidy is equivalent to the area T_1 A B T_2. Therefore, in a manner not dissimilar to that of economic monopolies which manipulate the supply curve to maintain artificial shortages in order to keep their prices high, ecological monopolies rely on the understatement of the real costs associated with environmental transformation.

THE PROBLEM OF ENVIRONMENTAL SHADOW COSTS

It is extremely difficult to assign a market value to nonmarketed goods, services, and life forms (see Munasinghe, 1993). Yet, without some means of price valuation, ecological interests will be consistently marginalized in relationship to more narrow financial objectives. Occasionally it is possible to utilize information from surrogate markets that implicitly provide some economic data about the way people value a nonmarketed good (e.g., the use of data in transportation markets to understand how people value their time in relation to the modes of transportation they select). However, such opportunities are all too rare.

With respect to ecological assets, the most common means of valuation is to calculate the value of an asset in terms of its value as an extractive resource in production processes or in terms of its value in situ with respect to recreational markets. In both respects, however, the fundamental criterion of valuation is strongly anthropocentric in character. A resource possesses economic value to the extent that it is useful to humans.

The issue of biodiversity poses especially difficult problems when it comes to economic valuations (see Tobey, 1993). How does one put a price tag on the loss of a plant or animal species? Given the incredible rate of species extinctions associated with human activities, one could only infer that economic decision makers assign a zero cost to the extinction of a species—or awfully close to that. Obviously, though, this cannot be true.

First, from a purely anthropocentric point of view, the irreversible loss of DNA in rare plants and animals places severe constraints on future genetic research. It is not inconceivable that tomorrow's cures for many of the dreaded diseases of humankind lie in the tropical forests that are being thoughtlessly bulldozed today. As biologist Daniel Janzen put it, "It's as though the nations of the world decided to burn their libraries without bothering to see what was in them" (Linden, 1989).

Second, intuitively one recognizes that the wealth of information contained in the DNA strands of rare plants and animals must have some intrinsic value. This information can never be retrieved once it is destroyed. It is not only part of the genetic heritage of life forms within the biosphere, it is also the raw material of all future discoveries in recombinant DNA research. The loss of such information logically has a price. Even if one were to value this information at the bargain basement price of only $0.50 per bit, the value of the DNA of endangered plants or animals would be staggering. For example, at $0.50 per bit, the information cost of the DNA of a frog species would range from $10 billion to $100 billion. Similarly, the information cost of the DNA would range from $7 billion to $70 billion for either a species of mosses or mollusks, from $9 billion to $900 billion for a fern species, and from $700 million to $5 trillion for a species of algae![5] Moreover, even in the unlikely possibility that the technological breakthroughs of the future will permit us to create DNA from scratch, one could surmise that the costs of fabricating DNA would be astronomical.

Ultimately, the problem of assigning shadow costs to environmental devastation stems from our inability to determine economic values apart from some type of market mechanism. The economic value of any good or service, for example, is not an attribute intrinsic to that good or service. The famous paradox of water and diamonds is a case in point. Water is a basic requirement of human life; its utility to humans is plainly evident. Diamonds, by contrast, may be aesthetically pleasing yet are not immediately useful to humans in any biological sense. Aside from a few industrial applications, diamonds are useful only as a means of adornment. Yet the economic value of diamonds is considerably higher than that of water. Why?

Adam Smith tried to resolve the paradox of water and diamonds by distinguishing between "value in use" and "value in exchange"; however, this distinction proved to be an artificial one, redescribing the paradox instead of resolving it. It was left to the theorists of the so-called Marginal Revolution (1871–1874)—most notably W. Stanley Jevons and Carl Menger—to offer a conceptual framework that resolved the famous paradox of value.[6] From this standpoint, the economic value of a particular good or service is not related to its overall usefulness (i.e., its total utility) but to the utility generated by consuming an additional *increment* of the good or service. Because water is generally a commodity in abundance, the consumption of an additional increment of water generates a small degree of additional utility relative to goods that are very scarce (e.g., diamonds). Accordingly the price of diamonds, a comparatively useless stone, is considerably higher than that of water when viewed from a marginal perspective. However, the *total* utility of water is greater than the total utility of diamonds. Having all the water in the world and no diamonds would be infinitely preferable to having all the diamonds in the world and no water.

The difficulty of assigning economic costs to environmental assets hearkens back to the water and diamonds paradox. Nearly all reasonable people of good conscience would affirm the overall worth of a healthy, clean environment. Clean air and unpolluted oceans, lakes, and rivers seem intrinsically good to most people in the late twentieth century. Moreover, most people, if faced with a choice between more biodiversity or less, would opt for preserving the rich diversity of our planet's biosphere. Yet, because environmental assets are so vast on the scale of mere human measures, they take on the appearance of being almost boundless assets.[7] This misperception of nature's infinitude feeds the illusion that the marginal destruction of environmental assets exacts few costs, because each increment of environmental damage is a comparative drop in the bucket in relation to the environmental assets that remain intact. From a marginal perspective, even the extinction of 100 plant and animal species per day can be rationalized by the fact that we have millions of species to go. The marginal worth of a single species in a sea of life is extremely low.

If environmental assets are to be valued in a rational manner, it is necessary to move beyond the valuational framework offered by marginal theory and standard supply and demand analysis. Otherwise such assets will be consistently undervalued because of the misperception of their abundance. Moreover, from a marginal perspective, one cannot fully appreciate the impact of discrete actions upon the whole because of the complex interdependence of the biosphere. Yet alternative valuational frameworks suffer from the problem of arbitrariness. Fortunately, a solution to the dilemma can be found by supplementing standard marginal analysis with a well-known criterion of economic efficiency known as Pareto optimality.

PARETO OPTIMALITY AND EFFICIENT HABITAT TRANSFORMATION

Pareto optimality is a concept used frequently by economists to identify efficient allocations of economic resources. To say that an allocation of goods and services is Pareto optimal means that you cannot make one person better off without making another person worse off—what game theorists call a zero-sum situation. The appealing feature of the Pareto criterion is that it is intended to prevent uncompensated, absolute losses for any party in a series of market transactions.

For example, let us assume that two societies—Paretio and Aparetio—have precisely the same demographic profile and the same natural endowments and economic resources. Moreover, let us assume that they have precisely the same gross national product (GNP) in a certain year. Given their exact same profile, one might expect that if the economy of Paretio satisfied the Pareto criteria, the economy of Aparetio must also be Pareto optimal. However, this may not be the case. For example, what if households within a lower income group in Aparetio suffered an absolute loss in income while other households experienced a gain in real income? If this loss was uncompensated by those groups that benefited from the economic production that year, Aparetio would *not* satisfy the Pareto criterion—even though its overall gross national product was exactly the same as the Paretio economy which did not violate the Pareto condition. Some households in Paretio's economy may indeed have experienced *relative* losses in relation to other households, but as long as other sectors of the economy did not gain at their expense, these relative inequalities are justified according to the Pareto criterion.

While Pareto optimality has acquired a kind of mantra-like status as a criterion of efficiency in mainstream economic theory, the concept is seriously flawed as a measure of allocative efficiency.

The primary difficulty facing the Pareto criterion concerns its indifference to the initial conditions of an economic distribution. The Pareto criterion can only determine whether there are Pareto improvements or violations over a specified time period. For example, let us conceive of a two-person society in which person A has an income of 10 consumption units and person B has an income of 200 units in year one. If, during year two, the income of person A remains unchanged and the income of person B grows to 225 consumption units, then this two-person society has satisfied the Pareto criterion, as the economic well-being of person

B increased without any deterioration of person A's income. Hence such an allocation of economic resources is deemed efficient, even though person A's income of only 10 consumption units could create tremendous suffering and hardship.

It is no accident that the Pareto criterion flaunts a studied indifference to the initial composition of an economic distribution. Since Lionel Robbins' influential *Essay on the Nature and Significance of Economic Science* (1932) most economists have assumed a constitutional skepticism about interpersonal comparisons of welfare.[8] By contrast, prior to the so-called ordinalist revolution of the 1930s, economists generally believed that it was possible to make interpersonal comparisons of welfare based upon an observable concept of utility. Generally speaking, they understood utility in relation to economic goods that promoted an individual's physical well-being in some sense (see Cooter and Rappoport, 1984). Without the possibility of interpersonal comparisons of welfare, the only type of welfare comparison that one can make is that a person is better off if he or she has an increased bundle of goods and services (i.e., more is better). Accordingly, the Pareto criterion is compelled to ignore how the composition of an initial distribution of economic resources may affect the efficiency outcomes of particular allocations of economic resources.

Another significant difficulty with Pareto optimality is that the criterion cannot be meaningfully applied to market economies without arbitrarily selecting an initial distribution. Since market distributions of goods and services represent an exceedingly complex series of economic transactions which simultaneously involve economic actors from several generations, it is impossible to select some nonarbitrary *initial* distribution from which the Pareto criterion can be applied. Without such an initial distribution, the concept is analytically meaningless. In the absence of such an initial benchmark, even Robin Hood could not be accused of violating Pareto optimality (Samuelson, 1947: 206).[9]

A further controversial feature of Pareto optimality that has been incorporated into modern cost-benefit analysis is the so-called Kaldor-Hicks principle of "hypothetical compensation" (see Sugden and Williams, 1978: 89–90; Little, 1979). This criterion holds that one economic reorganization is better than another if the winners could *hypothetically* compensate the losers. According to John Hicks,

> there is . . . a perfectly objective test which enables us to discriminate between those reorganizations which improve productive ef-

ficiency, and those which do not. If A is made so much better by the change that he could compensate B for his loss, and still have something left over, then the reorganization is an uneqivocal improvement. (1941: 108–16)

Shortly after Hicks and Kaldor proposed the hypothetical compensation principle, Tibor Scitovsky (1941) argued that, if compensation was hypothetical, the so-called Pareto improvements were reversible, in that the former distribution would be Pareto optimal in relation to the new distribution. Hence some type of actual compensation would be necessary in order to prevent this reversibility.

Despite the striking deficiencies of Pareto optimality as a criterion of efficiency for standard economic distributions, the concept is genuinely serviceable when it comes to ecological economics. This is true for four reasons.

First, because it is not necessary to think of ecological welfare from the standpoint of either cardinal or ordinal utility, environmental applications of Pareto optimality do not require interpersonal comparisons of welfare. While the consequences of a habitat transformation will impact its human inhabitants variously, depending upon their particular utility preferences and the extent to which environmental change affects them, it is possible to establish, on a scientific basis, standards of ecological well-being without resorting to the subjective and amorphous notion of utility.

Second, it is much easier to identify a credible ''initial'' situation in environmental issues, in contrast to the continuous flow of individual transactions that make up economic distributions. While habitats are continuously subject to the forces of change in a manner not dissimilar to that of market economies, the ''natural'' rate of change is considerably lower than in market economies. The dynamism of natural habitats is virtually undetectable to the casual observer, except when cataclysmic events impact the habitat (e.g., forest fires, volcanic eruptions, hurricanes). By contrast, change is a defining feature of market economies; one can neither ignore nor escape the dramatic dynamism (and habitat transformation) associated with contemporary economies. The pronounced differential in the rate of change between natural and man-made transformations of habitats makes it possible to identify initial benchmarks of ecological change against which Pareto improvements or violations can measured.

Third, when Pareto optimality is applied to environmental contexts, the Kaldor-Hicks principle of hypothetical compensation does not undermine the viability of the concept. Obviously, the environment cannot be "compensated" in an actual sense when it suffers a "loss" in a habitat transformation. It is not a reified entity that can be paid off or bribed to go along with a habitat transformation that will leave it worse off. The environment can be compensated as a losing party only in a hypothetical sense of the term. While a surtax on environmental monopoly profits could be both a helpful incentive for ecological responsibility and a source of government revenue for environmental protection, this cannot be viewed as compensation in any normal sense of the term. To do so would be to accept that the degradation of one habitat is somehow redressed by protective activities in another habitat.

Fourth, the Pareto criterion establishes a standard by which we can assess environmental costs from a supply-side perspective as opposed to a demand-side approach. As previously noted, the task of assigning shadow prices for environmental degradation has faltered on the intractable problem of estimating demand curves for environmental assets. Because we generally lack surrogate markets for assigning shadow prices to environmental assets, a demand-side approach to environmental loss will definitionally be highly arbitrary. This is to say nothing, of course, about the immense problems posed by the lack of intergenerational markets and the shortcomings of anthropocentric shadow pricing techniques (i.e., the value of environmental assets is their value to humans).

Roefie Heuting (1991) offers a practical solution to this problem by using a supply-side perspective. He argues that the social costs associated with environmental loss should be determined by the amount of money that would be required to produce a particular environmental outcome defined by society. While these standards for sustainability would necessarily be variable, given their dependence upon social choice procedures, such standards would give us the ability to understand environmental costs in terms of the costs of production associated with reducing environmental loss or maintaining an environmental standard. By using the Pareto criterion to establish an efficiency standard (which does not depend on highly variable social choice outcomes), it is possible to redefine environmental costs as the costs associated with restoring the environment to its original condition (i.e., hypothetically compensating the environment for the loss). Obviously, if these shadow production costs outweigh the anticipated financial returns of a project, it would be a poor investment.

In applying the criterion of Pareto optimality to ecological contexts

Table 5.1
Hypothetical Data on Habitat Transformations

Plan	Total Revenues	Rate of Return	Habitat Replacement Cost	Environmental Subsidy
A	$45 million	11%	$6 million	13.3%
B	$132 million	19%	$84 million	63.6%
C	$228 million	32%	$290 million	127.2%

one must begin with a defined habitat in a particular time period. Let us assume that three alternative types of habitat transformations (called plans A, B, and C) are being contemplated by governmental authorities (see Table 5.1). The extent of environmental transformation called for by Plans A, B, and C are of increasing magnitude. Plan A would have the least impact on the ecosystems of the target area, whereas Plan C would require dramatic environmental transformation. Each of the plans projects a different rate of return on investment over a 20-year period. Plan A has a projected annual rate of return of 11 percent, yielding total revenues of $45 million; Plan B has an annual rate of return of 19 percent, generating revenues of $132 million; and Plan C projects an impressive annual rate of return of 32 percent, yielding a total revenue of $228 million over a 20-year period. From the standpoint of the rate of return on investment, Plan C is obviously the best choice. But, is it more efficient than the other two?

We cannot answer this question until we calculate what the environment stands to lose in connection with plans A, B, and C. The easiest way to estimate this from an economic standpoint is to calculate how much it would cost to undo the planned habitat transformation. In effect, this amounts to estimating the "replacement costs" of a habitat prior to its transformation. Given the buoyant resiliency of nature, the costs associated with reversing most habitat transformations would be generally quite low.

Let us assume, for example, that Plan A has a projected habitat replacement cost of only $6 million, Plan B has a replacement cost of $84 million, and Plan C has a habitat replacement cost of $290 million. These habitat replacement costs are roughly equivalent to the "environmental subsidy" associated with the projected habitat transformation or the "monopoly profit" extracted by the human developers.

If one applied the Pareto criterion of efficiency under conditions of hypothetical compensation to Plans A, B, and C, it is clear that Plan C would be an unambiguous violation of the Pareto criterion and, hence, an inefficient habitat transformation. By contrast, both Plans A and B would comply with the Pareto criterion. From a purely financial stand-point, obviously Plan B is preferable, with its higher annual rate of return (19 percent as compared to Plan A's 11 percent). However, other con-siderations come into play that could strengthen the case for Plan A. For example, when one compares the habitat replacement costs (or environ-mental subsidy) to gross revenues, Plan A requires only a 13.3 percent environmental subsidy. By contrast, the subsidy for Plan B is a substan-tial 63.6 percent. If we used these percentages to adjust the rate of return on investment for Plans A and B, the unsubsidized annual rate of return on Plan A would be 9.5 percent as compared to 6.9 percent for Plan B.

Some may object that the condition of hypothetical compensation is too weak for the Pareto criterion to be a promising measure of efficiency for environmental transformation. One could imagine, for example, a project that is highly attractive from an investment point of view that relies on a whopping 99 percent environmental subsidy. Under the con-ditions of hypothetical compensation, such a project would not violate the Pareto criterion. While the government could tax the entire amount of the monopoly profit (or environmental subsidy) and channel these tax revenues into environmental protection projects, such stringent measures would presumably lack political feasibility. (However some tax schemes, such as a disincentive surtax on environmental monopoly profits, may be politically viable. This would require the development of environ-mental accounting to the point that environmental replacement costs could be calculated according to fairly objective criteria.)

Although the condition of hypothetical compensation might open the door to some abuse, it is clear that the Pareto criterion would go a long way in stemming the more pronounced forms of environmental exploi-tation in our day. Everything from dumping toxic wastes to destroying rain forests to depleting the ozone layer would qualify as plain violations of Pareto optimality under the conditions of hypothetical compensation.[10] While the Pareto criterion certainly does not preclude the exercise of monopoly power in habitats impacted by humans, it would certainly curtail the more destructive aspects of such monopoly power.

Change is an inevitable fact of life within any habitat. A guiding purpose of economic theory should be to find ways to channel change in a constructive manner. Indeed, one of the virtues of competitive mar-kets is that they are able to accommodate the dynamism of a changing economy without undermining its stability. The same attitude toward change and competition should carry over into the ecological sphere. Short of this, economic theory will continue to be strangely detached from its ecological moorings, proffering principles and prescriptions that undermine the well-being of humans and the health of the habitats in which they live.

EASY RIDERS AND FREE RIDERS

One of the most dramatic omissions of contemporary economic theory in relation to ecological issues concerns the so-called free rider problem. Where ecological issues are concerned, free riders are enjoined to be carefree, easy riders.

Standard fare, textbook treatments of the free rider problem link the concept exclusively with public goods. Public goods refer to forms of public investment that generate overall social benefits and cannot, by their nature, be restricted to those who paid for them. Unlike private goods, public goods are indivisible—possessing the capability to benefit parties who had no stake in paying for them. As such, public goods, for example, national defense and free education, generate positive "spill-over benefits" for society. Once these goods are paid for by taxpayers, individuals who have not paid for them can still benefit. One negative aspect of public goods, though, is that such goods allow some members of an economy to become free riders, living parasitically off the labor of others.

What is striking about the free rider problem in connection with economic theory is the complete absence of discussion as to how ecological monopolism promotes the proliferation of ecological free riders. Worse yet, there are several features of contemporary economic theory that legitimate and encourage people to become free riders in the economy. These factors include the following.

First, the concept of the long run in economic theory is extremely deficient when it comes to economic decision making in ecological settings—contexts that often require time horizons that span many generations, if not hundreds and thousands of years. The long run is typically defined as the period of time in which all inputs within a production process are variable. Consequently, the definition of the long run is highly telescoped in relation to biological time. Most often, though, the concept of long run functions as a mere formality in economic theory; the economic decisions that count from a business standpoint are usually constrained by a short-run time horizon. John Maynard Keynes' famous quote in his *Tract on Monetary Reform* (1923: 65) that "in the long run we are all dead" only underscores the contemporary irrelevance of long-run considerations to economic decision making. To state the obvious, this pronounced short- to middle-term time preference offers a powerful ideological justification for ecological free riders, as the daunting expanse of biological time is collapsed into a mere speck of time. The environmental impacts of economic policies must be manifest in a comparatively miniscule time frame if they are to be noticed at all.

Second, contemporary neoclassical theory is preoccupied with the way in which capital resources impact economic growth and development. While it is clear that one should not underestimate the importance of capital accumulation for robust and stable economic growth, the focus on physical capital has obscured the significant way in which both industrial and less developed economies rely upon substantial infusions of environmental capital in the day-to-day workings of their economies. This is particularly evident in the conspicuous absence of accounting for environmental depreciation in the conventional System of National Accounts (see Peskin, 1991). Moreover, the canonical status of present value analysis in the theory of finance has contributed significantly to both the myopic preoccupation with capital as a factor of production and the short-term perspective of economic policy.

Present value analysis relies upon the discount rate of *financial* capital in determining the worthiness of an investment; it does not take into account the considerably lower discount rate that would presumably apply to *environmental* capital. Because the discount rate reflects the time preference of a society for present consumption in relation to future consumption, it stands to reason that the discount rate for financial capital will always be higher than for environmental capital, as discussed previously. Therefore, to the extent that environmental capital is a significant component in an investment decision, present value analysis will discourage investment in projects that have longer time horizons and lower rates of return in the short term—presumably projects that are more likely to be environmentally friendly.

Third, mainstream economic theory evidences a strong bias in favor of market-based solutions to environmental problems which rely upon the clear definition of property rights (see Anderson and Leal, 1991). This rather pronounced bias stems from the highly influential theory of Ronald Coase (1960), known popularly as the Coase theorem. The Coase theorem holds that, if property rights are clearly defined and transaction costs are relatively low, social costs or externalities such as pollution can be efficiently regulated through private negotiations by what amounts to bribing behavior among the individual owners of property (i.e., compensating polluters as a means of altering their behavior). However, if property rights are not clearly assigned, resources held in common will be overexploited.[11] While the Coase theorem is sound in theory, it suffers from two major problems: (1) the high transaction costs implied by pervasive externalities and (2) the long-term insensitivity of market prices to resource depletion.

The first problem of the Coase theorem concerns what Daly and Cobb call "pervasive externalities" (1989: 55). Unlike the standard varieties of negative externalities, pervasive externalities generate external diseconomies that cannot be effectively localized (e.g., ozone depletion, deforestation, declining biodiversity, global warming). Hence market-based solutions, such as internalizing external costs, have little or no relevance for pervasive externalities.[12] Accordingly, where pervasive externalities are concerned, economic actors are in no position as individuals to prevent massive forms of environmental degradation by bribing polluters not to pollute. The transaction costs associated with such pervasive externalities are simply too great for the Coase theorem to be valid.

A second problem with the Coase theory concerns the short time horizons associated with market-based valuations of scarce resources. Such valuations are much too skewed in the direction of the short term to provide an adequate portrait of the intergenerational demand for resources. For example, Ezra Mishan (1979) has demonstrated conclusively that the Hotelling theorem (1931) is fatally flawed at its conceptual foundations. The Hotelling theorem holds that valuations of scarce resources in competitive markets will naturally lead to an optimal rate of exploitation of finite natural resources, as future shortages are supposedly anticipated by market prices and higher prices have a rationing impact on consumption behavior. Mishan, however, has shown that for the Hotelling theorem to work, it would require the torturous assumption of completely discrete generations (i.e., there could be no intergenerational overlap among economic decision makers). Consequently, while the Hotelling theorem is conceptually elegant on paper, it has little applicability to the major environmental problems we face.

The above features of contemporary economic theory have legitimated and encouraged free rider lifestyles among economic actors—a curious phenomenon for a discipline that is decisively hostile to free rider attitudes and economic forms of parasitism. One cannot help but conclude that economic theory is badly in need of a massive conceptual overhaul. Short of such reformation, the tools and concepts of mainstream economic theory will continue to subsidize all varieties of self-defeating ecological monopolies.

APPENDIX: EARTH CANCER AND THE UTILITARIAN TRADITION

The concept of utility, as it evolved within the utilitarian tradition, has experienced a progressive deterioration of meaning and has been pressed into the service of thoroughgoing anthropocentrism.

Originally, the concept of utility, as articulated by Jeremy Bentham, was directly related to the experience of pain and pleasure. The experience of pleasure increased utility; pain decreased it. Because Bentham believed that "pushpin is as good as poetry," all forms of human pleasure were ultimately undifferentiated and equal. In fact, Bentham allowed that the pain and pleasure experienced by all sentient creatures—not only humans—is morally relevant from the standpoint of utilitarianism. In his *Introduction to the Principles of Morals and Legislation*, Bentham wrote,

> The day may come when the rest of animal creation may acquire those rights which never could have been withholden from them but by the hand of tyranny. The French have already discovered that the blackness of the skin is no reason why a human being should be abandoned without redress to the caprice of a tormentor. It may one day come to be recognized that the number of the legs, the villosity of the skin, or the termination of the *os sacrum* are reasons equally insufficient for abandoning a sensitive being to the same fate. . . . The question is not, Can they *reason*? nor Can they *talk*? but, *Can they suffer*? (1789: 283n)

Significantly, the utilitarian tradition figured prominently in the path-breaking work in animal rights (e.g., Singer, 1975).

John Stuart Mill attempted to upgrade Bentham's undifferentiated notion of utility by distinguishing between lower and higher pleasures. This distinction, while rescuing the utilitarian tradition from the charge of being a vulgar and crude philosophy, had the effect of subverting the quantifiability of utility. Although Mill retained the Benthamite convention that it was possible to determine the aggregate utility of a whole by summing the individual utilities of its parts, the distinction between

higher and lower pleasures presumes that pleasure is a subjective quality—dependent on one's point of view—instead of a common property
attached to certain objects. Mill's subjective understanding of pleasure
is evidenced in the following passage from his well known essay, *Utilitarianism*:

> Neither pains nor pleasures are homogeneous, and pain is always
> heterogeneous with pleasure. What is there to decide whether a
> particular pleasure is worth purchasing at the cost of a particular
> pain, except the feelings and judgment of the experienced? (1861:
> 15)

Mill's introduction of a strong subjective orientation in the theory of
utility had the effect of making utility synonymous with "desiredness,"
that is, *human* desiredness. This effectively cut off any possibility of
utilizing the principle of utility to weigh the interests of sentient life
forms in a policy context. More important, the subjective understanding
of utility that Mill introduced later became very influential in the way
in which such economists as Alfred Marshall and A. C. Pigou defined
utility in connection with economic theory (Little, 1957: 20).

Today the term "utility" is simply shorthand for the notion of "revealed preferences." According to contemporary consumer theory, when
people reveal their preferences by actually choosing one product over
another, they are necessarily maximizing their utility. However, this ultimately amounts to the vacuous claim that whatever people choose max-
imizes their utility. Consequently, the theory of revealed preferences
solidly welds the link between utility and human preferences. Is it impossible to define utility apart from the actual demonstration of individual preferences? Moreover, the progressive demise of the notion of
cardinal utility introduced substantial doubts about the plausibility of
interpersonal comparisons of welfare. In particular, with the development
of the concept of Pareto optimality as the standard measure of allocative
efficiency, skepticism about interpersonal comparisons of welfare
achieved a kind of canonical status within economic theory.

The net effect of the evolution of the concept of utility in the history
of utilitarianism is the progressive identification between the social op-
timum and the satisfaction of individual preferences. This, in turn, provides a powerful justification for earth cancer, as nearly all contemporary
notions of utility know no limits to the unbridled expression of human
preferences—from either the standpoint of economic efficiency or morality. Economics, sadly, cast its lot with utilitarian thought early in its

disciplinary history and, today, shares the impoverishment of the utilitarian tradition.

NOTES

1. The primacy of Ricardo's thought in the evolution of economic thinking receives eloquent confirmation by the way in which both neoclassical and Marxist theorists have self-consciously appropriated his ideas.

2. It may be objected that voluntary exchange is an integral characteristic of markets; one cannot have a market in any sense of the term without it. While it is clear that reference to "ecological markets" is both nontraditional and represents a considerable extension of the normal contexts of the term "markets," the application is by no means unjustified. Voluntariness, as a quality adhering to markets, is by no means a universal quality. Monopolistic or monopsonistic markets, for example, seriously diminish the voluntariness of both consumers and producers; yet they are no less markets because of this trait.

3. One could imagine, of course, that most species, if they had a choice, would favor fine-tuned environmental transformations that would make them more effective at predation, parasitism, or colonialization. Yet, given the immensely intricate interdependencies of the natural world and the relative dearth of open genetic programs (Mayr, 1970, 1976) that permit "adaptive" or "intelligent" behaviors, it seems reasonable to assume that the collective preference of nonhuman life forms would be biased against change.

4. Shadow costs refer to costs that cannot be accounted for in traditional market values (e.g., the cost associated with the extinction of a plant or animal species). As such, they are inputted costs that are assigned according to a particular valuation methodology.

5. Estimates are drawn from Raff and Kaufman (1983: 314), calculated at only two bits per nucleotide pair (i.e., it is only necessary to read one nucleotide to deduce the composition of the pair).

6. While the concept of marginal utility, as applied to the paradox of water and diamonds, provided the most adequate framework for resolving the paradox, in point of historical fact the paradox had been resolved several times previous to Jevons' *Theory of Political Economy* (1871) and Menger's *Principles of Economics* (1871). The most notable resolution was that of the Italian economist Ferdinando Galiani (*Della Moneta*, 1751). See Schumpeter (1954: 300ff) for a full discussion.

7. Even some economists have asserted that natural resources are virtually boundless in light of the unfolding horizon of technological possibility. For example, Julian Simon contends that continued population growth poses little problem with respect to the utilization of scarce natural resources as "such resources are created by mankind in response to human needs" (1977: 484).

8. During the 1930s, Robbins and others within the academic community

were strongly influenced by the positivist ideas that were being formulated in Vienna. He dismissed a "cardinal" notion of utility on the basis that it was not empirically verifiable. Moreover, he defined utility strictly in terms of subjective preferences and correctly noted the impossibility of making interpersonal comparisons on this basis. This association between utility and subjective preferences diverged sharply from the notion of utility employed by the economists from the material welfare school. (Some economists prior to Robbins understood utility in a more subjective sense, such as Jevons, 1911: 14.) Hence the former notion of cardinal utility was replaced by a subjective, "ordinal" notion of utility which held that one could identify whether an individual's relative utility had increased or decreased only by measuring additions or losses to an individual's bundle of economic goods.

9. Because there is a multiplicity of Pareto optimal distributions, each corresponding to a particular initial distribution, a Robin Hood–style redistribution would itself generate a new Pareto optimal distribution (assuming the workings of a competitive market economy), even though it was predicated upon a violation of the Pareto criterion. Hence, without establishing an initial distribution, there is no way to determine whether the criterion is being satisfied or violated.

10. I assume that the extraction of minerals and fossil fuels would only be factored into an estimate of habitat replacement costs inasmuch as such extractive enterprises impact other life forms. Presumably there are no moral obligations to refrain from exploiting inanimate objects, except as this impacts other life forms and future generations.

11. A corollary to the Coase theorem is Garrett Hardin's (1968) "tragedy of the commons." Hardin argues in a persuasive fashion that resources that are held in common will always be overexploited by individual economic actors.

12. For an alternative point of view, see Anderson and Leal (1991: 154ff). For example, they note the potential use of tracer techniques (e.g., adding odorants, coloring agents, isotopes to air pollutants) as a means of "fencing the atmosphere" (i.e., defining property rights of air space in order to internalize external costs through market-based solutions).

Chapter 6

Discovering Our Humanity

FOUR BIOTIC DECISIONS

Humans pride themselves on being creatures of choice. Appropriately so. For humans, the capacity to choose creates new horizons of both possibility and predicament. The ability to conceive and reflect upon alternative courses of action is a hallmark of humanity. As Ernst Mayr has noted, the development of "an ethics based on decision making was perhaps the most important step in humanization" (1988: 77). Moreover, it is in the act of choosing that we define ourselves—irrespective of whether such choices are processed with complete self-awareness or reinstated by the force of habit and culture.

Biological life, for humans, entails four basic types of decisions: (1) survival decisions, (2) reproductive decisions, (3) predation decisions, and (4) habitat decisions. Unlike most other life forms on the planet, we enjoy the luxury of having a considerable range of control over these decisions. They are decisions thrust upon us neither by dint of instinct nor predators. While culture is certainly a key factor in how these decisions are made and reinstated, we should not forget that humans create culture in a significant respect by choosing to reinforce or undermine cultural norms (see Giddens, 1984). Facing these decisions responsibly is a birthright of being an intelligent life form. To do otherwise is to relinquish our humanity.

Humanity is not a quality we can define in isolation from our species life within the biosphere. It is a quality that emerges in the decisions we make. Our humanity evokes humility not because human nature is a synonym for weakness but because our humanity provokes us to find our place in the universe—and this cannot help but be humbling. The dignity made possible by our humanity is not a quality that adheres to us biologically. That dignity, or in the words of Camus "common decency," is either diminished or reaffirmed by the decisions we make. The anguish of plague calls on us to decide who we really are, and in this our faculties for dignity are discovered.

Survival Decisions

Survival decisions are focused on the preservation and perpetuation of individual life. To a great extent we do not recognize these as decisions per se. Instead we routinely tend to discount the voluntariness of these decisions—either by attributing the desire for survival to an "instinct" for self-preservation or by establishing powerful social sanctions that bias these decisions in one way or another. Perhaps we discount the voluntary character of these decisions because of the mystery that surrounds them. Take, for example, two types of survival decisions: suicide and self-annihilating altruism.

Suicide represents an explicit decision about individual survival. In its most fundamental form, suicide is an acute response to hopelessness—whether this springs from profound depression, a terminal illness, personal tragedy, or a loss of meaning. Because survival is assumed to be good until proven otherwise, the decision to take one's life requires some form of internal justification. As such, the institution of suicide, at one and the same time, affirms the human preference for life and the human need to live with hope and meaning.

The human preference for life has a kind of axiomatic quality. Yet, since we are creatures who thrive on intangibles such as hope and meaning, life defined in the absence of hope and meaning loses its power to captivate us. In this regard we may have something in common with the whales that appear, on occasion, to commit a kind of mysterious mass suicide—for purposes that evade us—by intentionally beaching themselves in great numbers.

Another type of survival decision is the decision to sacrifice one's life altruistically for the sake of another person or an ideal. This type of self-annihilating decision is often directed at those individuals who are genetically related to the altruist. The rather widespread manifestations of sacrificial behavior throughout the animal kingdom—what sociobiolo-

gists call "hard-core altruism" or "kin selection"—seem to be closely
related to reproduction. An altruistic animal that sacrifices its life in order
that a close relative may live thereby enhances the survival of its own
genes, in spite of the fact that the individual organism has sacrificed its
own reproductive capability by the altruistic act (Hamilton, 1964; Wilson, 1975: 117–20; 1978: 155–56).

That said, among humans, hard-core altruism appears to take on another dimension, due to the impact of mythic and theoretic consciousness
on human behavior. Hard-core altruism, in such cases, becomes a vehicle
for demonstrating certain spiritual or social ideals, such as love or loy-
alty. The saint or hero who performs a self-sacrificial act displays a
willingness to dispense with biological existence for the sake of some
higher good. A similar dynamic holds true for most expressions of pa-
triotism (see Niebuhr, 1932), where the survival of the community (and
the values/ideals it represents) takes precedence over the survival of the
individual. These well-attested displays of altruism are probably related
to Abraham Maslow's (1948) observation that people tend to place a
greater value on higher "self-actualization" needs than on lower needs
when they "have been chronically gratified in both."

Perhaps the most remarkable aspect of the human will to survive is
the persistence of the "human spirit." Many people who suffer crushing
adversity characteristically remark that they found the strength to survive. What at a distance seems forlorn and intolerable becomes bearable.
Is the will to survive that strong within the human breast? Perhaps. However, one would imagine that the persistence of the human spirit has
more to do with human capacity to hope.

Hope and one's degree of investment in the future are integrally
linked. For many people, children, a next generation, become the focus
of hope. Others focus their hope around the achievement of certain goals
or some anticipated change in their life situation. Whatever the case,
hope seems to be a critical ingredient for human determination. The
challenge that humans face at this point in history is to discover sources
of hope that are not immediately dependent upon reproduction or perpetually increasing consumption. Indeed, the hope we need is of a religious character—one that appreciates the limitations of human finitude
without compromising our responsibilities to preserve the biosphere.

Reproductive Decisions

Whereas reproductive decisions are "instinctual" for virtually all life
forms that inhabit our planet, humans at the end of the twentieth century

enjoy an unparalleled degree of choice in their reproductive decisions. Yet, even though we have the luxury of consciously making reproductive decisions, one cannot help but be forced to the conclusion that we have made such decisions irresponsibly.

Each year the world's population increases by more than 90 million people. This amounts to adding a city the size of Anchorage, Alaska, or Reno, Nevada, or Lincoln, Nebraska, or Savanna, Georgia, to the world's population every day! Altogether this is equivalent to adding the combined populations of Denmark, Finland, the Netherlands, Norway, Sweden, and the United Kingdom to the planet each year (Brown, Flavin, and Postel, 1991: 21).

Nearly all of this new population growth (90 percent) will take place in less developed countries. In the next 40 years the population of Sub-Saharan Africa will soar from 500 million to 1.5 billion. Asia's population will swell from 3.1 billion to 5.1 billion, and Latin America's population will rise from 450 million to 650 million. The collective impact of this pattern of intensive and concentrated population growth on the environment will be devastating. For example, today only a small segment of humans live in habitats that exceed population densities of 400 people per square kilometer (e.g., Hong Kong, Singapore, South Korea, Bangladesh, the island of Java, Indonesia). However, by the middle of the next century, one-third of Earth's human population will live in densely populated habitats that exceed 400 people per square kilometer. In Bangladesh alone the population densities will explode to 1,700 people per square kilometer (World Bank, 1992: 7).

Clearly something must be done. As of yet, though, the response of the international community has been tepid, at best, or has been characterized by denial and utter negligence, at worst. The 1992 Rio Earth Summit, for example, gave virtually no attention to the problem of curbing population growth (*The Economist*, 1992a). Such an omission is stunning. It is an issue fraught with political snares—not only for the governmental regimes of less developed countries, but also within the industrialized world. To make matters worse, some of the global institutions that are positioned to play a constructive role in facilitating a concerted response to this issue are passionate advocates for inaction (e.g., the Catholic Church). Our inability to take responsibility for a biotic decision as basic as reproduction is a profound commentary on human irresponsibility. It provokes the disturbing question, recently posed by Harvard biologist E. O. Wilson (1993), "Is humanity suicidal?" Will we eventually suffocate under the weight of our numbers, transforming a bountiful planet into a diminutive prison cell?

A first step in confronting this problem is to take responsibility for our reproductive decisions. They are not decisions of human caprice, fate, culture, or God—they are our decisions. In this regard, it is critical to increase the sphere of influence of women in making reproductive decisions. This can be done in several ways, but certainly the most effective among them is to give women the power to read. More schooling translates into more economic opportunities and fewer children (Brown, Flavin, and Postel, 1991: 106). A woman who has the knowledge and self-esteem to make her own decisions about reproduction will more likely make reproductive decisions that serve the best interests of her family and society as a whole. The World Fertility Survey of 1980 and other studies have shown that approximately one-half of all married women do not want more children. Most, however, do not use any form of birth control (Mann, 1992). The remarkable convergence between the women's movement and the environmental movement on this issue is a tremendous resource in confronting the global challenge of overpopulation.

Should decisions about family size be safeguarded as a fundamental human right? I would surmise that an overwhelming majority of Earth's human population would respond affirmatively to this question if a global survey were conducted on this issue. Most people would strongly resent the intrusion of government into such matters. Yet it is difficult to see how there can be anything like a right of reproduction, particularly when such decisions place the Earth's ecology at such extreme peril.

That said, it seems both inadvisable and infeasible for most governments to play a direct role in reproductive decisions within the household. It is considerably more promising for governments to play an indirect role by using public policy to promote responsible reproductive decision making. This can be encouraged in two ways: (1) by empowering women to take control of their own reproductive decisions and (2) by increasing the opportunity costs of having children.

The first goal relies primarily on education. It requires an increase in traditional and nontraditional educational opportunities for females, as well as focused contraceptive education (including maximal access and control over the means of contraception). Throughout most of the world, decisions about family size usually have more to do with the culture-based expectations of a wife's mother-in-law and her husband's concept of virility than with her own reproductive choices. Answering the question—"What makes a good wife?"—typically has more to do with the quantity of children she produces than the quality of care she extends to herself and her family. Education offers women an opportunity to gain the confidence to listen to their own voices in such matters. In addition, the empowerment of women is directly tied to the demise of cultural practices that provide an economic rationale for the subordinate social status of women. Both the custom of the parents paying a dowry for their daughters and the practice of suitors paying a bride price for their prospective wives (which thereby purchases their labor) create a strong economic disincentive for parents to invest in their daughter's education (Brown, Flavin, and Postel, 1991: 106–7).

The second goal requires initiatives designed to increase the opportunity costs of having children by enhancing the earning potential of women in the marketplace. The concept of opportunity costs, in economics, refers to the cost of foregone opportunities. The decision to bear and raise children is a significant time investment that entails substantial tradeoffs. These tradeoffs only become apparent, though, when female decision makers are presented with other investment opportunities for their time and effort. By giving women the option of investing their labor into marketable products and services, the opportunity costs of having children are increased. Hence women opt for having fewer children (Cochrane, 1979; World Bank, 1984: 109–11).

In the final analysis, what Sarah Ruddick (1989) calls "maternal thinking" is undoubtedly our strongest asset in confronting the problem of overpopulation. Presumably no woman really wants to be treated as a baby farm. Moreover, the relational dimension of unfettered maternal thinking naturally places a high premium on quality of life issues that impact the household. If there is anything like a maternal instinct, it is most pronounced in decisions to place the interests of the family first. For example, in household consumption surveys a consistent pattern has emerged: Women who have some discretionary control over household income (generally by earning money through their own enterprises) will be more apt to spend it on basic goods that benefit the family as a whole than their husbands. The empowerment of women simultaneously advances the power of maternal thinking in human relations, and this cannot help but be in the long-term interests of mother Earth.

Predation Decisions

Humans are not carnivores by nature, but learned carnivores (Crook, 1980: 148). Yet it is impossible to speak of the development of human social life without also acknowledging the central role that hunting has played in human development. Most of the traits that are associated with human sociality proceeded from a revolutionary new cultural development known as the hunting and gathering way of life. The basic structure of this new form of social organization emerged perhaps as early as five million years ago, but the phenomenon of large-scale, big-game hunting (and the high level of social cooperation/communication it required) was a far more recent phenomenon, developing within the past two million years (Leakey and Lewin, 1977: 148).

It is likely that the sociological innovation of the hunting and gathering way of life, perhaps in connection with the development of weapons and tools, was a primary factor in the remarkable growth of the human neocortex, as a result of the increased demands for communication and social interaction associated with big-game hunting (Crook, 1980: 147–48; Eibl-Eibesfeldt, 1989: 610–11; Isaac, 1983). The hunting and gathering way of life was typified by central-place food sharing, a division of labor between males and females, extensive geographic mobility, the increased use of tools, and a high degree of intragroup coordination (Crook, 1980: 127–34). Obviously, the discovery of fire and the innovation of cooking food (which certainly was widely practiced 500,000 years ago, but may be as ancient as 2.5 million years ago) markedly enhanced the sociality of hominids by extending the day and providing a central place for social interaction (Leakey and Lewin, 1977: 131).

The feeling of community that hunter societies, such as Native American tribes (see Chapter 2), evidence in connection with the animals they hunt finds poignant expression in Ice Age cave paintings. The fact that these paintings were located in remote chambers that are virtually inaccessible suggests that the paintings had some ceremonial purpose, instead of being created as decorative art for one's living quarters. Take, for example, the cave paintings at Niaux, in the French Pyrenees. One has to walk through a kilometer of pitch-black corridors before reaching a damp, chilly, circular hall that features large images of ibex, bison, and other hunted animals. Only with the aid of a flickering fire could Niaux's earliest visitors discern the marvelous creatures that inhabit its inner chamber (Donald, 1991: 281).

Given the central position that hunting has occupied in the development of human sociality (and probably human intelligence), it is peculiar that the environmental movement displays a bias against hunting and, to a lesser degree, against meat consumption. Predation is an ecological fact of life. The biospheric destruction wrought by humans in the contemporary world is primarily the consequence of habitat degradation caused by human population growth, patterns of settlement, the generation of wastes, and resource exploitation—not predation. Nonetheless there are areas where predation decisions become significant from the standpoint of ecological interdependence. For example, the problem of overharvesting the world's oceans poses an increasing threat to both life in the sea and the human settlements that depend upon the fishing industry for their livelihood. Since 1950 the annual marine fish catch has quadrupled worldwide, reaching 84.2 million metric tons in 1990. At present 8 out of the world's 17 ocean fisheries are being exploited beyond the minimum range of the estimated sustainable catch with 3 of those 8 exceeding the maximum range—the Southeast Pacific, the Northwest Pacific, and the Mediterranean and Black seas. In New England alone the overfishing of cod, haddock, and flounder has lead to the lowest catch of commercially valuable fish on record—entailing a loss of 14,000 jobs and $350 million in income (World Resources Institute, 1992: 7, 180).

While the criterion of sustainability should be the primary consideration in predation decisions, it may not be the only consideration. For example, the use of factory farming to mass produce animal flesh for commercial distribution satisfies the criterion of sustainability in terms of the preservation of a species. Yet, the appalling conditions of these animal prisons, where life consists of interminable feeding on death row, not only deny animals the semblance of a natural life but also deny humans of their humanity. One simply cannot respect the value of ecological interdependence and, at the same time, subject helpless animals to the worst possible living conditions during their meager lives—all in the name of production efficiency.

Similarly, from the standpoint of sustainability, the controlled harvesting of the minke whales and Norway's defection from the ban on commercial whaling by the International Whaling Commission (IWC) seems justified. There are an estimated 900,000 minke whales worldwide. In light of their relative abundance, the scientific committee of the IWC determined that a limited hunt would not endanger the species (Lemonick, 1993a). Yet, is it always the case that species which *can* be harvested, on grounds of sustainability, *should* be harvested? One could point to two other considerations that may constrain predation decisions significantly. For ease of reference, I will refer to the first consideration as the "ecological justice constraint" and the second as the "close relative constraint."

Can we speak of ecological interdependence apart from some embryonic concept of ecological justice? Does a history of abusive predation introduce special obligations, from one species to the next, that attempt to redress a past injustice, in much the same way as systems of justice operate in human communities? Specifically, in the case of whaling, do we owe whales anything in light of their bleak history of exploitation at human hands? The grisly legacy of human whaling surely cannot be glossed over in this regard. For example, in Antarctica's South Ocean alone, fifty years of whaling reduced the populations of humpback whales to just 3 percent of their original population and blue whales to 5 percent (World Resources Institute, 1990: 192). What conceivable justification can be found for such ecological downsizing? Whereas no justice can be found in the life and death competition characterized by the "law of the jungle," whaling, for all but a few Arctic peoples, has never been a life and death matter of subsistence but a matter of commercial gain.

The fact that whale populations may never recover completely from human exploitation, owing to low rates of reproduction, warrants extreme caution in connection with any controlled hunting proposal. A case in point is Antartica's blue whales. Even though they have not been hunted for nearly 30 years, their population still lingers at only about 500. Worse yet, after sixty years of a hunting ban on the northern right whale, there are still only a maximum of 350 right whales that grace the Atlantic (Lemonick, 1993a).

The ecological justice constraint requires us to come to terms with a history of exploitation in a predation decision. Although minke whales themselves have not been hunted to the brink of extinction, largely because their smaller size made them less commercially attractive, should we not afford all whale species the opportunity to thrive in the Earth's oceans? Do we not owe all whales the opportunity to be free from the exhausting chase of a high-tech whaling boat and the unforgiving shock of a piercing harpoon? Those who dismiss such a position as being sentimental and unscientific should themselves provide a coherent rationale as to why our history with whales should be deemed irrelevant in the present context.

Furthermore, one could argue from a strategic point of view that a partial lifting of the international ban on whaling could erode the strength of the ban in protecting truly endangered whale species. Because international law derives its legal weight largely from the force of custom, the concept of a selective ban might make sense from a scientific standpoint, yet lack durability in the wild and woolly world of sovereign nation-states. Moreover, there are always ingenious ways to cheat on quotas, as attempts to control the trade of ivory have shown, making the notion of globally controlled hunting a chimera. Therefore it seems neither callous nor capricious to deny commercial whalers the opportunity to harvest these magnificent creatures, even when we can find a scientific rationale for it.

A second constraint on predation decisions might be termed the ''close relative constraint.'' This constraint holds that intelligent species have a prima facie obligation to avoid predatory relationships with other intelligent life forms. Unlike the ecological justice constraint, this predation constraint is based on respect for intelligence instead of respect for biospheric interdependence. In some respects, the close relative constraint represents the flip side of cognition conceit (see Chapter 3). Whereas cognition conceit understands intelligence in such a way that it reduces biospheric relationships to global ''I-it'' relationships, treating intelligence as a singular human attribute, the close relative constraint treats intelligence as a foundation for ''I-Thou'' relationships with other species. Although it may seem unusual to refer to human-animal interaction in terms of I-Thou relationships, consider E. O. Wilson's reflections upon his encounter with the pygmy chimpanzee named Kanzi, discussed in Chapter 2.

Kanzi walked into the office and spotted me sitting on a chair on the far side of the room. He went into a frenzy of excitement, yelping and gesticulating to the two women with him in a way that seemed to exclaim, ''That's a stranger! Why is he here? What are we going to do about him?'' . . .

The trainer gave him a cup of grape juice, and he climbed into her lap to drink it and be cuddled. After a short wait he slid down to the floor and drifted back over to me. . . . I imitated the flutelike conciliatory call of the species, wu-wu-wu-wu-wu. . . . Kanzi reached out and touched my hand, nervously but gently, and stepped back a short distance to study me once again. The trainer gave me a cup of grape juice of my own. I flourished the cup as if offering a toast and took a sip, whereupon Kanzi climbed into my lap, took the cup, and drank most of the juice. Then we cuddled.

Afterward everyone in the room had a good time playing ball and a game of chase with Kanzi.

The episode was unnerving. It wasn't the same as making friends with the neighbor's dog. I had to ask myself: was this really an animal? As Kanzi was led away (no farewells), I realized that I had responded to him almost exactly as I would to a two-year-old child—same initial anxieties, same urge to communicate and please, same gestures and food-sharing ritual. Even the conciliatory call was not very far off from the sounds adults make to comfort an infant. (1984: 128–29)

The cumulative effect of encounters such as this one is that the line between human and animal is blurred and a foundation for relatedness is built. This makes predatory actions unseemly at best, and immoral at worst.

Returning to the debate concerning minke whales, from the standpoint of the close relative constraint, the unqualified ban on commercial whaling is justified because of the presumptively high degree of intelligence exhibited by whales. There is simply too much complexity in whale behavior—from exquisite mating calls to seemingly mass suicides—not to presume that they are intelligent creatures. The same holds true for dolphins. The international movement to ban the use of driftnets in fishing for tuna and other species (which also entangle and kill hundreds of thousands of dolphins) is a humane response to the plight of a "close relative" (World Resources Institute, 1992: 181).

Habitat Decisions

Humans are unique in the way in which they make decisions about habitats. They not only make decisions about the habitats they colonize but also, through waste generation and resource exploitation, make decisions that have far-reaching impacts on habitats they barely see, much less occupy. Unfortunately most of the current environmental crisis is attributable to this distinctive human capacity—magnified exponentially by uncontrolled population growth through human reproductive decisions.

The history of human settlements provides elaborate testimony of the need for territorial possession among humans. Whether one's habitat is an apartment in midtown Manhattan or a squatter settlement in Mexico City or a Bushman hut in the Kalahari desert, human dwellings not only provide people with a modicum of privacy but also are an important medium for individual expression. As Irenäus Eibl-Eibesfeldt observes,

> The home in the simplest case consists of a windscreen, providing protection against the elements. It marks the place of residency and allows for privacy, a need that can be demonstrated universally. Even where several families build a communal house, each family has its own sleeping area and fireplace. It does not require great effort to provide in basic terms for privacy in a community and also to maintain protection from the elements and predators. The small round hut of the Bushmen fulfills this need as does the lean-to shelter of the Yanomami or the windscreen of the Agta. . . .
>
> The poor of the tropic metropolitan areas often live on the edge of the city in apparently crowded conditions with poor hygienic facilities, but each family has its own house, constructed perhaps of cardboard, sheet iron, and boards, and each has neighbors who are usually friends and relatives. The children also have contact with others, playing perhaps in the dirt but with their friends and in the open air. Each tiny house has a little yard in which melons, vegetables, and flowers are grown in rusty cans and in which chickens are free to run about. This is a more stimulating environment for young and old than a cell in a high-rise structure which may be hygienic but in which one lives in social isolation and where the sun rarely shines. (1989: 631, 638)

The need for territorial possession among humans is evidenced in a myriad of ways, ranging from students who always sit in the same seat in a classroom to families that always vacation in the same place each year to city dwellers who desire to own property in the countryside or along coastlines (Eibl-Eibesfelt, 1989: 630). Moreover, this human need is amplified significantly by cultures that have well-defined notions of private property, where the acquisition of resources becomes an expression of territoriality (see Sahlins, 1972: 92–94; Wilson, 1978: 109–10). Consequently the magnitude of one's possessions becomes the measure of social status, leading to an ever-expanding definition of individual habitats and a widening search for new territory and the resources to support it. There would be little economic incentive to cut down rain forests if the demand for construction materials was more modest. Obviously, the population explosion only feeds the consumptive appetite for escalating territorial possession. Nowhere is this more evident than in the contemporary phenomenon of the city.

In 1950 there were two metropolitan areas in the world with a population above 8 million: London and New York City. By the year 2000 there will be 28 such metropolitan areas above 8 million (United Nations, 1991: 23). The combined metropolitan populations of Mexico City and São Paulo alone are nearly four times the size of New York City in 1950. In fifteen short years, the urban population within the Third World will double. Given this explosive urban phenomenon, nearly all of which is taking place in the Third World, planned and controlled urban growth is virtually impossible. For example, for the past decade Bangkok has ingested 3,200 hectares of surrounding farmland to feed its appetite for growth—an area equivalent to more than one-half the size of Manhattan (Worldwatch Institute, 1992: 119, 134).

The voracious hunger of the massive metropolis for food, water, and energy has placed tremendous pressure on outlying farmlands and aquifers while contributing to environmental degradation through air and water pollution. In the countryside, overcultivation, deforestation, and desertification have conspired to disintegrate the basic resources for our survival. Every year we sacrifice 24 billion tons of topsoil from Earth's croplands because of erosion wrought by human hands. This is equivalent to scraping off one inch of topsoil from 50 percent of China's arable land and dumping it into the Pacific Ocean (Worldwatch Institute, 1993: 12). Each year! In Africa the forces of overpopulation, overgrazing, price distortions, deforestation, and desertification have conspired to deny the continent the ability to feed itself. In 1990 Africa produced less than 80 percent of what it needed in food grains. The World Bank estimates that Africa's need for imported food will double in 2010 (Stackhouse, 1993).

The impact of human habitat decisions upon the rest of Earth's ecology is both immense and immensely disturbing. Thousands of species go extinct each year as a direct result of the decisions that humans make about their own habitats and those that lie beyond the fringes of human settlement (McNeely et al., 1990: 12, 43–74, 56–62; World Resources Institute, 1994: 149–51). Like a cancer that threatens to devour a healthy organism, we are making self-defeating biotic decisions that reduce humanity to being a plague upon the planet, instead of being the crown of Earth's creation. It is most disturbing that we will never fully know how many opportunities have been forever lost because of thoughtless habitat decisions. Take, for example, the case of Jalisco maize (*Zea diploperennis*). When it was discovered in the 1970s south of Guadalajara,

Mexico, by a college student, the species was only one week away from extinction by machete and fire. This disease-resistant species is unique among contemporary maize species in that it possesses the capacity for perennial growth. If its genes can be transferred to domestic corn, it would dramatically increase the global output of corn. And it was one week from extinction (Wilson, 1992: 281).

Among the more remote yet perilous effects of human habitat decisions is the decay of coral reef systems. The Great Wall of China is the largest structure on the planet built by human hands. Yet it is not the largest structure built by Earth's inhabitants. That distinction goes to the tiny coral polyps that have built the exquisitely beautiful reef systems that decorate the ocean's floor—a marvel only scuba divers can fully enjoy. Although coral reefs cover considerably less than 1 percent of the ocean floor (0.17 percent), it is estimated that one-fourth of all ocean species find their homes among coral reefs. They are the rain forests of the sea.

Unfortunately, these marvelous creatures are highly vulnerable to relatively minute changes in seasonal ocean temperatures of only one or two degrees above the maximum seasonal range. For example, when the severe El Niño of 1982–1983 raised ocean temperatures in the eastern Pacific above the seasonal average by three or four degrees, it is estimated that between 70 percent and 90 percent of the corals in Panama and Costa Rica died and that more than 95 percent of the corals of the Galápagos Islands perished. Unless revived, such reef systems are destined to disintegrate into the ocean. Obviously, the prospect of ocean warming because of urban-concentrated CO_2 pollution does not bode well for the survival of coral reefs and the multitude of species they support.

The problem of ocean warming is further compounded by ocean pollution and sedimentation due to logging, soil erosion, and mining. Sedimentation blocks out precious sunlight from which the coral polyps derive their energy and forces the corals to devote what remains of their energy to cleansing the reef. For example, in Bacuit Bay of the Philippines coastal logging increased soil erosion by more than 200 times. As a result, 5 percent of the corals in the bay perished in 1985—after only one year of logging.

Furthermore, human predation decisions compound the adverse effects of these habitat decisions. In this regard, overfishing among reefs is a persistent problem. The depletion of fish stocks often leads to the invasion of sea urchins which undermine the structure of reefs. For example, in Kenya, the overharvesting of two species of triggerfish was followed by an invasion of sea urchins on its coastal reefs. As a consequence, the densities of these reef-eating urchins reached as high as 65 per square meter. Worse yet, absurd fishing techniques that employ the use of explosives or cyanide further degrade reef systems. In the Philippines alone it is estimated that 30 percent of its reef systems are largely dead. Another 39 percent of the reef systems had only 25 to 50 percent covering of healthy corals. Only 5 percent of the Philippine's magnificent reefs were in pristine condition!

The plight of coral reefs around the world is a foreboding illustration of how human habitat decisions (and, secondarily, predation decisions) have far-reaching implications for habitats that only a few visit and most of us hardly understand. We cannot take lightly the essential interdependence of our biosphere and our responsibility to respect life in all its diversity. The most complex of human achievements cannot begin to rival the intricate edifices built by corals. Awe, not apathy, is the proper human response (Brown and Ogden, 1993; Worldwatch Institute, 1993: 42ff).

EMBRACING THE K-TRANSITION

The portrait that emerges from these four biotic decisions—especially reproductive and habitat decisions—is one of humanity out of control. The facts could not be more plain and unequivocal. We are not making responsible biotic decisions and, thereby, we are unwitting agents of earth cancer. Much of our current predicament can be explained by what I call the K-transition (Weigel, 1989: 217–26). The K-transition refers to the passage between cultural forms that evolved in circumstances of expanding and variable habitats (r-cultures) and cultural systems that are suited to stable habitats at or near their ecological carrying capacity (K-cultures).

The distinction between r- and K-selection was first developed by MacArthur and Wilson (1967) as a model of density-dependent natural selection in connection with their work on the colonization of islands.[1] Where natural selection favors high population growth, as in the case of an initial period of colonization, r-selection or r-strategies become the most influential (''r'' is the symbol used to signify the rate of population increase). By contrast, K-selection becomes prominent when the habitat of a species is at or near its ecological carrying capacity (''K'' represents the carrying capacity of the habitat). In K-conditions, natural selection

Table 6.1
Comparisons between r-Selection and K-Selection

Characteristic	r-Selection	K-Selection
Climate	Variable	Constant
Availability of Resources	Unpredicatable	Predictable
Habitat	Transient	Stable
Mortality	High	Low
Competition (both within a species and between species)	Weak (unimportant)	Strong (important)
Energy Utilization	Emphasizes Quantity	Emphasizes Quality
Population Size	Varies Dramatically	Steady (close to carrying capacity)
Age at Reproduction	Younger	Older
Number of Offspring	Many	Few
Parental Care	Poorly Developed	Well-Developed
Social Grouping	Poorly Integrated	Well-Integrated
Altruism	Rare	Common

Source: Modified from David Barash, *Sociobiology and Behavior* (New York: Elsevier, 1977), p. 183.

favors K-strategists who utilize resources more efficiently and allocate parental investment increasingly for rearing offspring instead of reproduction. In short, an r-strategy for reproduction emphasizes the quantitative dimensions of life by producing as many offspring as possible, whereas a K-reproductive strategy focuses on increasing the quality of life by limiting the birthrate. Table 6.1 compares the major differences between r-selection and K-selection.

For humans, the distinction between r-strategies and K-strategies are inextricably connected to the process of cultural evolution. Cultural evolution, as distinguished from genetic and cognitive evolution, has played a decisive role in recent human development. Whereas the pace of genetic evolution is glacial and that of cognitive evolution is slow (see Chapter 2), cultural evolution is a comparatively rapid process that is shaped by the emergence of new ideas which interact with the envi-

Table 6.2
r-Cultures vs. K-Cultures for Humans

r-Cultures	K-Cultures
uncontrolled population growth	controlled population growth
poorly integrated social groupings	well-integrated social groupings
unmanaged economies	managed economies
altruism rare	altruism common
weak or nonexistent regulatory mechanisms for governing intra- and inter-specific competition for resources	strong regulatory mechanisms for governing intra- and inter-specific competition for resources
energy utilization extensive	energy utilization intensive

ronment in much the same way that gene mutations open up new possibilities for genetic evolution through natural selection (see Dawkins, 1976, 1982).[2] While nonhuman animals have the capacity to *transmit* culture to other members of their species,[3] the human ability to *accumulate* (and hence evolve) culture makes the process of cultural evolution possible (Tinbergen, 1973: 201).

In light of the pervasive impact of cultural evolution upon human behavior, it is reasonable to speak of r-cultures and K-cultures, referring to cultural forms that embody either r- or K-reproductive strategies. Table 6.2 presents a suggestive list of the traits associated with r-cultures and K-cultures in human communities.

The most prominent feature of K-cultures is their moderate to high degree of regulation. The laissez-faire character of r-cultures and their social institutions is conspicuously absent. The regulatory landscape of K-cultures is explained, in part, by the keen competition for resources that usually characterizes K-selection. For humans, such competition could either encourage cultures that reward unmitigated narcissism, as in the case of the Ik culture in northern Uganda (Turnbull, 1972), or cultures that utilize a broad repertoire of mechanisms for environmental regulation. Since the practice of unmitigated egoism is usually socially destabilizing and diminishes our adaptive capabilities, we are left with the second alternative: some degree of regulation for governing the competition over scarce resources. Without effective environmental regulation, r-cultures will continue to dominate human behavior while we are rapidly approaching K-conditions for our species. This is our present situation.

In this regard, it is interesting to note that r-culture fantasies about the colonization of space or Earth's oceans (as a response to K-conditions) represent both the denial and the affirmation of the inevitability of increasingly regulated environments. On the one hand, such fantasies represent the denial of K-conditions by perpetuating the myth of expanding habitats and geographic frontiers. On the other hand, the technological requirements of such exotic habitats presume a high degree of intragroup coordination and regulation, signaling a new set of constraints upon human freedom and behavior.

The troubling dimension of the greater degree of regulatory mechanisms within K-cultures concerns the extent to which such cultures will be able to preserve a genuine sense of individual feedom within the inherent constraints of bureaucratic oversight and management. Certainly totalitarianism of both the left and right fails miserably in providing effective environmental regulation because of the lack of effective checks and balances within the social structure (e.g., China's one-child family policy is very environmentally friendly; its industrial waste policies for air and water pollution are not). Hopefully we can find imaginative solutions for environmental regulation that will create a system of checks and balances that allows for a meaningful sphere of individual choice while achieving the aims of preserving our planet's ecology.

The passage from r-cultures to K-cultures—the K-transition—represents an especially difficult and unstable phase of human cultural evolution. It will not be easy, particularly in light of the capacity for denial and self-deception among humans. There will always be those who protest even the very notion of our planet's carrying capacity, believing that God, technological progress, or sheer luck will rescue us from the prospect of ecological tribulation. We must not lose hope, though, that humanity will rise to the occasion. The consequences of despair and inaction are simply too great. Perhaps we can learn a lesson from the Maya of Central America.

Since the discovery of Mayan ruins in the Honduran jungle during the mid-1800s, the mysterious remnants of this majestic civilization have lured archaeologists, anthropologists, and linguists from around the world. As the fascination with the Maya grows, the more we know about this sophisticated civilization.

Where did the Maya come from? We simply do not know. A consensus has emerged, though, that by 900 B.C. the civilization had spread across the region we now know as Mexico's Yucatan Peninsula, Belize, and the northern half of Guatemala. Between A.D. 250 and A.D. 900 the ancient Maya civilization reached its zenith. The achievements of the Maya include the development of the most elaborate writing system of the Americas, a sophisticated command of mathematics, and a remarkably accurate astronomical calendar. They also built massive pyramids across Central America that have become modern-day monuments to their cultural legacy.

One of the great riddles of history is the sudden disappearance of Maya culture. The civilization vanished within 100 short years of the beginning of the eighth century. While we cannot be sure about what led to the demise of the Maya, it appears that a tradition of incessant warfare and increasing elitism conspired with the environmental problems of overpopulation and the overexploitation of the rain forest ecosystem. Whether the requirements of war and the building projects of Mayan elites precipitated the civilization's downfall by placing increasing demands on the environment, or whether overpopulation and increasingly scarce resources provoked a recurring state of war, we simply cannot determine at this time. It is clear, however, that the Maya overexploited their environment. The data from pollen found among underground debris indicate that most of the tropical forest had been destroyed. Moreover, it appears that the Maya had population densities of 200 people per square kilometer—a figure that rivals the most densely populated areas of the world prior to the Industrial Revolution (Lemonick, 1993b).

The lesson for us from the Maya, according to Vanderbilt archaeologist Arthur Demarest, is that "any society that depends on growth economics, with elites demanding ever-greater levels of material well-being, eventually reaches its limits" (Weisman, 1990: 42). For Demarest the civilization was caught in a spiral of escalating consumption driven by the society's elites and their penchant for dynastic warfare. The more temples that had to be built, the more food had to be supplied to the workers. The need for increased food production correspondingly required the need for more people. Also, the demands of unrelenting war likely contributed to increased population growth, thereby expanding pressures on the environment. This counter-intuitive relationship between war and population growth is evidenced in the Aztec, Incan, and Chinese civilizations, where constant warfare stimulates population growth in response to the demand for soldiers (Weisman, 1990).

If Demarest and other experts are correct, the consumptive demands of the Mayan civilization exceeded the carrying capacity of their environment. In effect they were an r-culture that failed to negotiate the transition from an r-culture to a K-culture. Hopefully the lessons of the Maya will not be lost on us. The social, economic, and political challenges posed by the K-transition are daunting, yet not unsurmountable. More will be said about this in the next chapter.

There are also critical, identity-based questions raised by the K-transition. One cannot approach human biotic decisions responsibly without also asking fundamental questions about who we are as humans. Two of those pivotal questions are (1) Is human life sacred? and (2) Is there such a thing as a *human* right? The remainder of this chapter is devoted to a brief consideration of these two questions.

THE SANCTITY OF HUMAN LIFE RECONSIDERED

Is human life sacred? Does the sanctity of human life diminish the sanctity of other life forms? Is human life sacred in a way that other life forms are not? These and other questions are essentially religious in character. They are also questions with tremendous ethical import.

The concept of the sanctity of human life has been very influential in ethics. Not only is the notion strongly rooted in the tradition of theological ethics, but also themes relating to the sanctity of human life have been prominent within philosophical ethics. For example, early contract theories of morality built upon an implicit notion of human sanctity by employing the concepts of inalienable rights (Locke) or inalienable freedoms (Rousseau). Moreover, in significant respects, the Enlightment notion of autonomous rationality—a keystone concept in both the Kantian (e.g., Donagan, 1977) and libertarian (e.g., Nozick, 1974) traditions of ethics—can be understood as a secularized version of the sanctity of human life.

The thought of Immanuel Kant is indicative of the influential role that the sanctity concept has played in philosophical ethics, particularly in the way in which he invokes the notions of personhood and dignity. In Kant's famous *Groundwork of the Metaphysic of Morals* (1785), he defends his second formulation of the categorical imperative[4] by reference to the concept of personhood:

> Now I say that man, and in general every rational being, *exists* as an end in himself, *not merely as a means* for arbitrary use by this or that will: he must in all his actions, whether they are directed to himself or to other rational beings, always be viewed *at the same*

time as an end. . . . Beings whose existence depends, not on our will, but on nature, have none the less, if they are non-rational beings, only a relative value as means and are consequently called *things*. Rational beings, on the other hand, are called *persons* because their nature already marks them out as ends in themselves—that is, as something which ought not to be used merely as a means—and consequently imposes to that extent a limit on all arbitrary treatment of them (and this is an object of reverence). (1785: 64–65)

Similarly the notion of dignity figures prominently in Kant's defense of the third formulation of the categorical imperative. In this third formulation, Kant defines morality in terms of actions that are consistent with what he calls a "kingdom of ends." He describes this kingdom as "a systematic union of different rational beings under common laws," and later he distinguishes between things which have a "price" and those which have "dignity":

In the kingdom of ends everything has either a *price* or a *dignity*. If it has a price, something else can be put in its place as an *equivalent*; if it is exalted above all price and so admits of no equivalent, then it has a dignity. . . .

Now morality is the only condition under which a rational being can be an end in himself; for only through this is it possible to be a law-making member in a kingdom of ends. Therefore morality, and humanity so far as it is capable of morality, is the only thing which has dignity. (77)

While the sanctity of human life has been influential in philosophical ethics, the foundations of the concept belong to theological ethics. Within the Judeo-Christian tradition, the concept of the sanctity of human life is integrally tied to the conviction that humans were created in God's own image. In the Christian tradition, the theme of *imago dei* has been an especially prominent feature of Christian ethics. Genesis 1:26–27 is the *locus classicus* of both faiths:

Then God said, "Let us make humankind in our image, according to our likeness; and let them have dominion over the fish of the sea, and over the birds of the air, and over the cattle, and over all the wild animals of the earth, and over every creeping thing that creeps upon the earth.'' So God created humankind in his image, in the image of God he created them; male and female he created them.[5]

There have been a number of interpretations of Genesis 1:26–27, focused on what it means to be created in the image of God. These include, but are not limited to, the following:

1. The text affirms the intrinsic worth of humans in God's eyes (e.g., in the *Enuma Elish*, the Babylonian cosmogony that is closest to the Genesis account, humans were created from the blood of a rebel god for the purpose of relieving the gods of tiresome work).
2. The text refers to the attributes that humans share in common with God (e.g., the capacity for moral choice, freedom, love, compassion).
3. The text refers to the capacity of humans to be creative.
4. The text refers to the priority of humans over the rest of created order (i.e., humans are God's viceroys).
5. The text refers to the idea that humans were created with both the capacity and longing to know God.
6. The text refers to the idea that humans were created with a soul.
7. The text refers to the idea that all humans have a portion of God within them.

Sometimes Genesis 2:7—''then the Lord God formed man from the dust of the ground, and breathed into his nostrils the breath of life; and the man became a living being''—has been linked to the last two interpretations of Genesis 1:26–27. This linkage has been based upon a misinterpretation of the Hebrew word *nephesh*, which is sometimes translated ''soul'' but refers to the totality of one's being (hence, ''man became a living being'').

Which of the above interpretations, if any, accurately depicts the significance of humans being created in the image of God? Presumably the meaning of Genesis 1:26–27 cannot be determined without reference to the entire creation narrative in which the passage is set.

What is striking about the first creation narrative (Gen. 1–2:4a) is the intimate portrayal of God's relationship with creation. Genesis 1:26–27 is located within a creation account that places considerable emphasis on God's care in calling forth the sun and stars, the sky, the oceans, vegetation, fish, birds, wild animals, and "living creatures of every kind." Each time that God called forth life, whether fish or sea monster, "God saw that it was good." It was only on the sixth day of creation that God created humans. Then, on the sixth day,

> God blessed them, and God said to them, "Be fruitful and multiply, and fill the earth and subdue it; and have dominion over the fish of the sea and over the birds of the air and over every living thing that moves upon the earth." God said, "See, I have given you every plant yielding seed that is upon the face of the earth, and every tree with seed in its fruit; you shall have them for food. And to every beast of the earth, and to every bird of the air, and to everything that creeps on the earth, everything that has the breath of life, I have given every green plant for food." And it was so. (Gen. 1:28–30)

Whereas many recent critics of this passage have understood it as sanctioning an imperialistic posture toward nature, particularly focusing on the harsh word "subdue" (Hebrew *cabash*) and the term "dominion" (Hebrew *radah*), what is remarkable about this passage is its global and throughgoing commitment to vegetarianism. The human prerogatives to "fill the earth and subdue it" and to "have dominion over the fish of the sea and over the birds of the air" did not include killing them for food. The birthright of all living things is to be herbivores; predation upon "everything that creeps on the earth" is ruled out by implication— not only ruled out for humans but also for all other creatures.[6]

In Genesis' second creation account (Gen. 2:4b–25), this image of complete peace among all creatures and our connection with nature receives dramatic affirmation. God sets the first human in the garden of Eden "to till and keep it" (Gen. 2:15), not for the purpose of developing or subduing it in the normal sense of the term. Moreover, God made it clear that the first human would derive his sustenance from the trees within the garden (Gen. 2:9). Most remarkably, God, being concerned about the first human's aloneness, sought out friends for him among the existing community of animal life (Gen. 2:18–20). In so doing, God took special care to orient the first human to all life forms, bringing all creatures to him so that he could name them. It was only when he could not find a companion suitable for the first human among the community of animals that God elected to create another human. Significantly, when the prophet Isaiah described the messianic kingdom, he used the images of the wolf and lamb living together peacefully and the leopard sleeping alongside a young goat (Isa. 11:6–9), harkening back to the comprehensive *shalom* that characterized God's original creation (see Oelschlaeger, 1994: 136–37).

What does it mean that humans were created in the image of God? Perhaps the most natural meaning of the text (Gen. 1:26–27) is that humans were created to see creation the same way that God sees it, to relate to it as God relates to it. The mandate to "subdue" and "have dominion" is at one and the same time a mandate to keep, to preserve, and to cherish, celebrating our community with other life forms. If this means that humans are God's viceroys on Earth, they are "gardener-kings" who take tremendous pride in the wonder and beauty of the natural world, not marauding overlords who see the natural world as something to be transformed or "improved."

The portrait of unabridged tranquility in God's original creation contrasts sharply with the strife and discontent after the Fall of humans (Genesis 3). The first sign of human's spiritual alienation from God is the invention of clothing—a sign of human physical alienation from nature. In response to the Fall, God intensified this sense of alienation from nature putting "enmity" between the serpent and humans,[7] increasing the pangs of childbirth, and reducing the fertility of the soil, so that "in toil you shall eat of it all the days of your life" (Gen. 3:17b).

Certainly the most dramatic manifestation of this alienation was God's decision to destroy all life on earth with a flood, saving only Noah and his family along with representatives of the animal kingdom (Gen. 6:11–8:19). Consequently, all life forms suffered as a result of the alienation between God and humans. Following the flood, in what may be the most remarkable passage of the Bible, God expressed regret for destroying animal life on the account of humans: "I will never again curse the ground because of humankind, for the inclination of the human heart is evil from youth; nor will I ever again destroy every living creature as I have done" (Gen. 8:21).

In looking at the creation accounts of Genesis, we find virtually no support for an anthropocentric orienation to life. Instead, human life is integrally related to biological life on the planet. Humans are created in God's image to assume the vocation of gardening, not engineering. Gardeners understand the intricate interdependence of nature; engineers build pretentious towers of Babel to defeat nature. The sacredness of human life is rooted in God's affirmation of the sacredness of all life.

In sharp contrast to the Hellenistic equation of sacredness with separateness, an assumption that informs most contemporary concepts of the sanctity of human life, Hebrew thought was profoundly holistic in character. Sacredness describes God's estimate of the goodness of all life; it adds to without taking away. The force of *imago dei* is not to exalt the priority and prerogatives of humans over nonhuman life, but to affirm the profound interconnectedness between humans and all other life forms. The sanctity of human life is also the sanctity of all life.

HUMAN RIGHTS RECONSIDERED

The concept of human rights is perhaps the most powerful moral idea in contemporary ethical theory. The belief that humans have justifiable rights claims that adhere to every person—regardless of social status, sex, ethnicity, or nationality—provides a robust conceptual framework for moral obligation. Yet the notion of human rights also bears witness to the anthropocentric fallacy in ethics when it is used as a vehicle for exalting the powers and prerogatives of humans over other species.

One can construct a foundation for human rights from several ethical traditions, including natural law theory (Maritain, 1944, 1951), libertarianism (Nozick, 1974), social contract theory (Rawls, 1971; Shue, 1980), and Kantianism (Donagan, 1977; O'Neill, 1986). However, it is likely that the theory of human rights developed by Alan Gewirth (1978, 1979, 1982) provides the firmest, most compelling basis for the human rights concept. Not only does Gewirth's theory represent the most rigorously argued moral theory to date, but also the theory has recently received further refinement and a meticulous, elaborate defense by Deryck Beyleveld (1991).

For Gewirth the starting point of moral reflection is action or, more precisely, human agency. Action is to moral philosophy what empirical data are to the natural sciences. Building on a minimalistic notion of human rationality (i.e., people are able to trace logical entailments and avoid self-contradiction), Gewirth argues that all humans are logically constrained to affirm—on prudential grounds—both their right to the necessary requirements of human agency and the rights of others to these universal requirements. He defines these necessary requirements, or "generic features" of action, in terms of freedom and well-being.

The key to Gewirth's argument is what he calls a "dialectically necessary method" (1978: 44). This dialectical process mimics, in a Socratic fashion, an internal conversation that begins with the person's recognition of his or her own prudential claim for the requirements of human agency (i.e., freedom and well-being) and concludes with the moral judgment that all persons have rights to these basic goods. In essence, we are able to enter the agent's own perspective and to translate an agent's informal, practical judgments into a series of logically entailed statements.

The genius of Gewirth's dialectically necessary method is that it produces an external and universal principle of morality from the prudentially motivated, internal reflection of the agent. Gewirth calls this universal moral principle the Principle of Generic Consistency (PGC): "Act in accord with the generic rights of your recipients as well as yourself" (1978: 135). Working solely from the prudential interests of the agent, Gewirth demonstrates that an individual cannot logically affirm his or her claim to the necessary requirements of human action without granting the same right to others.

While there are alternative ways to formulate Gewirth's argument for the PGC (see Beyleveld, 1991), the argument originally presented by Gewirth is as follows. First, a person begins by recognizing that all purposive action is valuational (agents act for purposes they construe as being good, where "good" is a value-neutral concept and simply refers to a pro-attitude of the agent toward his or her actions) and that he or she requires freedom and well-being as necessary preconditions of action. Hence, when an agent (actually or prospectively) claims (1) "I do X for purpose E" (as all choosing, purposive agents must), it follows that (2a) "E is good" and (2b) "My freedom and well-being are necessary goods." Because the agent necessarily views his or her own purposes as good (2a) and recognizes that certain goods are requisite preconditions for the realization of these purposes (2b), it logically follows that (3) "I have rights to freedom and well-being." Gewirth interprets the movement from statement 2b to statement 3 as simply reflecting the agent's prudentially motivated realization that any purposive action requires freedom and well-being. No significance should be attached to the rights language that Gewirth uses here. In fact, as Beyleveld argues, Gewirth's concept of dialectical necessity requires that all judgments of value are prudential in character (1991: 38ff). Hence statement 3 is founded on the structure of human action instead of proceeding from an exogenous theory of rights.

Once the veracity of statement 3 is established, Gewirth demonstrates that statement 3 logically entails (4) "All other persons ought at least to refrain from interfering with my freedom and well-being.'' Gewirth justifies this move on the basis of the formal structure of rights claims (''*A* has a right to *X* against *B* by virtue of *Y*'' [65ff]), which locates individual rights claims in a correlative relationship with one's moral obligations to others.

Having established the logical entailments from statement 1 to statement 4, Gewirth introduces what he calls the Logical Principle of Universalizability (i.e., ''If some predicate *P* belongs to some subject *S* because *S* has the property *Q* [where the 'because' is that of sufficient reason or condition], then *P* must also belong to all other subjects S_1, S_2, . . . S_n that have *Q*'' [105ff]). This Logical Principle of Universalizability, in turn, yields what Gewirth calls the Criterion of Relevant Similarities (''If one person *S* has a certain right because he has quality *Q* . . . , then all persons who have *Q* must have such a right''). Once the truth of the Criterion of Relevant Similarities is accepted, Gewirth proceeds to demonstrate that statement 4 entails (5) ''All prospective agents have rights to freedom and well-being'' because being a prospective purposive agent is the sufficient condition (*Q*) for the possession of generic rights (*P*). From statement 5, it logically follows that (6) ''I ought at least to refrain from interfering with the freedom and well-being of any prospective purposive agent.''

Finally, in light of the transactional structure of action (which involves agents as well as recipients of their actions) and the fact that often negative moral obligations logically imply the existence of positive moral obligations (where an agent ought to assist his recipients in having ''freedom and well-being whenever they cannot otherwise have these necessary goods and he can help them at no comparable cost to himself'' [135]), it follows from statement 6 that (7) ''I ought to respect the freedom and well-being of any (and every) prospective purposive agent.'' The imperative form of statement 7 is the PGC: ''Act in accord with the generic rights of your recipients as well as yourself.''

The structure and logic of Gewirth's theory seems impeccable. Although I have argued elsewhere (Weigel, 1989: 75ff) that significant problems are associated with Gewirth's understanding of freedom and well-being as the requisite requirements of action and have proposed an alternative formulation based on the concept of basic human needs, these issues do not detract from the philosophical integrity of Gewirth's theory. Yet, appropriately, one may ask whether human agency deserves to be the point of departure of moral reflection through Gewirth's dialectically necessary method. Is this decision mandated by logic and rationality or is it a reflection of the profound influence of the anthropocentric fallacy in ethical theory?

One suspects that the decision to make *human* agency the cornerstone of ethical theory has nothing to do with either logic or ratiocination. Instead the PGC exemplifies the anthropocentric fallacy in that it confuses the human capacity for reflective, purposive action with the aims of ethical theory. For example, the PGC offers no protection whatsoever to the nonhuman recipients of human agency (except, perhaps, as the welfare of nonhuman *recipients* tangentially impacts the well-being of humans). On what basis should we exclude nonhuman life forms as being morally significant recipients of human agency? The question is especially compelling in light of the massive loss of biodiversity that is the direct consequence of human agency. One could imagine many ways in which a theory of human rights that is based on human agency could mask gross forms of cancerous colonization under the noble aegis of morality and human rights.

In my book, *A Unified Theory of Global Development* (1989), I tried to address this issue by broadening the focus of the PGC beyond the requirements of human agency. Using the concept of basic human needs, I argued that morality should be founded not on the structure of human agency per se, but on the essential characteristics or "species universals" of human life. As such, human *being* is a more adequate basis for morality than human *agency*. Building on the three species universals of human existence, human intelligence, and human sociality, I constructed a theory of basic human needs that addressed the major problems facing Gewirth's concept of generic goods (i.e., freedom and well-being). More important, this modification permitted a reinterpretation of Gewirth's theory as a "species ethic," based on what I referred to as "the open genetic program of human life," which found its ultimate justification in the "goodness" of environmental adaptation. Utilizing Gewirth's dialectically necessary method, I derived two principles of morality similar to the PGC: (1) the Basic Needs Mandate: All human beings have the right to meet their basic needs; and (2) the Basic Needs Imperative: Act in accord with the basic needs of other human beings as well as yourself.

As species ethics, I argued that the Basic Needs Mandate and the Basic Needs Imperative would expressly prohibit policies or actions that would lead to the extinction of other plant or animal species. I made the argument on the ground that it would be logically inconsistent for any species ethic to sanction courses of action that would lead to the extinction of other species (1989: 80–81). Not only would this deny the essential interdependence of our biosphere, but it would require the justification of supraspecific criteria that endorse the existence of certain species and legitimate the extinction of others. I noted that the only possible supraspecific criterion that would qualify is that of evolutionary complexity (i.e., more complex [higher] life forms have a stronger claim to survival than less complex [lower] life forms). For example, if the very existence of a small insect threatened the survival of an endangered primate population, it would be permissible to destroy the insect species on the basis of the criterion of evolutionary complexity. The same would be true in connection with viruses that pose significant threats to higher life forms, such as the destruction of the remaining vials of the smallpox virus. However, such acute interspecific problems are extremely remote. Moreover, the criterion of evolutionary complexity is an especially precarious one—not only because of the difficulties involved in defining and measuring complexity, but also because the criterion could theoretically justify the extinction of the human species by more advanced extraterrestrial life forms. Therefore, in the absence of definitive supraspecific criteria, a theory of human rights that is conceived as a species ethic must necessarily respect the biological integrity of other species.

While I believe that my reinterpretation of Gewirth's theory protects it from legitimating gross forms of biospheric abuse, both the Basic Needs Mandate and the Basic Needs Imperative are still anthropocentric principles. Indeed, one would imagine that any theory of human rights will necessarily be based on an anthropocentric principle of one type or another. Yet, a failing of both Gewirth's theory and my own reformulation of the PGC as a species ethic is the tacit assumption that a theory of human rights is largely coterminous with what we can know about universal moral concepts.[8] While neither the PGC nor the Basic Needs Imperative are restricted in application solely to human beings, both principles require a baseline of rationality before life forms are deemed "morally significant."[9] In this sense, both principles are guilty of committing the anthropocentric fallacy.

The force of this discussion is that the concept of human rights must not be construed in an absolutist and LIFE-denying manner. The human rights concept is an important vehicle for instilling an increased level of respect for persons. Yet notions of human rights, like ideas about the sanctity of human life, lack authenticity if they do not also magnify our respect for all life forms. A human right that denies the integrity of LIFE is a biologically disembodied notion. To live well as humans and to realize fully our humanity is to grasp our place in the world, experiencing the passion and wonder of life . . . that of our own lives, that of the lives of other human beings, and that of the cohort of life that we call the biosphere . . . on the planet named Earth.

NOTES

1. The terms "r" and "K" are derived from the logistic equation $dN/dt = rN(K-N)/K$, where N is the population size, r is the intrinsic rate of increase, and K is the carrying capacity of the habitat.

2. Zoologist Richard Dawkins emphasizes this point by introducing the concept of "memes," referring to idea replicators that have an incredible impact in shaping the course of human cultural evolution. Memes, for Dawkins, are to cultural evolution what genes are to genetic evolution.

3. For some interesting examples of cultural transfer among animals, see Wilson (1975: 168–72).

4. The second formulation of the categorical imperative is "Act in such a way that you always treat humanity, whether in your own person or in the person of any other, never simply as a means, but always at the same time as an end."

5. From the New revised Standard Version. All subsequent quotations from the Bible are from this version.

6. The first explicit reference to the killing of animals for *food* is God's covenant with Noah (Gen. 9:3). The parallelism with Genesis 1:30 is obvious.

7. See Wilson for a fascinating discussion on the biological and cultural significance of serpents for human communities (1984: 83–101).

8. For example, Gewirth refers to the PGC as "the supreme principle of morality" (1978: 145), and I characterized the Basic Needs Imperative as "a supreme moral principle by virtue of its universality, not supreme in the sense that it evokes the highest level of moral conduct in humanity (i.e., as a religious ethic of sacrificial love)" (1989: 85). In both instances, the PGC and the Basic Needs Imperative are understood as providing an Archimedean point by which all other ethical principles, moral ideals/virtues, social policies, and cultural systems can be evaluated.

9. Technically, the PGC would apply to all life forms capable of reflective, purposive agency and able to recognize self-contradiction and to trace logical entailments. Similarly, the Basic Needs Imperative would apply to all life forms who exhibit characteristics/capabilities that are universal to the species and facilitate environmental adaptation, as well as the ability to recognize self-contradiction and to trace logical entailments.

Chapter 7

Passion for Life

FAITH AND THEORETICAL CONSCIOUSNESS

Cogito, ergo credo—"I think, therefore I believe."

Since the dawn of human consciousness, humans have been extraordinarily curious about LIFE on Earth.[1] From primitive fertility goddesses to sophisticated world religions, *Homo sapiens* has been integrally occupied with questions about the nature of LIFE—the purpose of their own lives, the origin of human life, and the origin of LIFE itself. While such queries are often expressed in the quest to get behind or above LIFE in order to discover the true meaning of individual life, they are first and foremost questions about LIFE itself . . . prompted by the inherent quandaries of human life. The complexity and magnitude of LIFE provoke the need to transcend LIFE in order to understand life.

The Neanderthal burial rituals of the Upper Pleistocene may suggest the beginnings of our attempts to fathom the mysteries of LIFE. Approximately 60,000 years ago, a Neanderthal man was laid to rest on a bed of flowers in a place we now call the Shanidar cave of Iraq's Zagros Mountains (Leakey and Lewin, 1977: 177). The vast gulf of time that separates us from those who performed the burial ritual makes it impossible for us to determine their intent. We can only infer that death for them, as for most of us, was an experience cloaked in mystery and awe.

The embrace of death holds a dilemma for humans. Individual life ceases while everything else seems to endure. This apparent contradiction between the finitude of individual life and the apparent infinity of time and space (as experienced through biospheric LIFE) has been a central theme of religious experience (Giddens, 1979: 161). Presumably the developing capacities of humans for self-consciousness and abstraction made it increasingly difficult to ignore the incongruities that death poses.

One of the earliest attempted resolutions of this contradiction was the ancient phenomenon identified by the late Mircea Eliade (1954, 1959) as "the myth of the eternal return." This refers to the belief, held by many ancient or nonliterate peoples, that time could be mythically reversed by repeating the ritual recreation of the world. As such, the myth of the eternal return provided humans with the ability to place the infinity of time and space under human control, giving humans the prerogative to recreate or repeat historical processes whenever they choose. Through the ritual creation of time, humans could be the masters of time, not its victims. The same is true for the more elaborate notions of afterlife that developed later, where the contradiction between finite life and the apparent infinity of time and space was resolved by extending finite life into infinity.

Human theoretical consciousness faces a similar dilemma in our day. That contradiction revolves not so much around the mystery of death as around the meaning of individual endeavor within the vast sea of biological time. It is a contradiction born from the diminutive posture of human significance, dwarfed against the backdrop of Earth's dramatic landscape of LIFE. On the one hand, humans, if they are to find their place in the world, must relinquish their position of preeminence—risking a conception of life in order to find life. On the other hand, finding one's place in the world seems to condemn human endeavor to insignificance, as our achievements and aspirations are puny in comparison to the grand designs of nature.

To find life one must lose life. This paradoxical truth is a central theme of religious experience. Yet while human mythical consciousness has celebrated and enshrined this theme, it is a theme rarely explored by human theoretical consciousness. Perhaps this is because the experience of faith seems to be at odds with the analytical ambience of theoretic culture. The constrictive boundaries of the scientific method cannot comprehend the convictions of faith. "I think" has too often been translated as "I observe," to the detriment of "I experience" or "I believe." Yet, it is in the realm of conviction and faith that the individual grasps his or her relationship to the whole. Understandings of "totality" can be constructed only with the building blocks of knowledge, belief, and experience. More important, the act of faith—demonstrated by the expression of hope—is an integral aspect of penetrating the whole. For an individual to have faith is to penetrate the vastness of the whole, affirming at one and the same time the connection between "I am" and "all that is." To have passion for LIFE is to affirm that connection. Without this it will be impossible to resist the malignacy that threatens to engulf our planet with cancer.

THE BUDDHIST PARADOX

Buddhism, as a religion and a philosophy, is preoccupied with grasping the significance of whole. For Gautama Buddha and those who followed, the pathway of individual liberation is virtually synonymous with the process of losing oneself in the vastness of the whole. As such, the role of the individual in the world at large is to be free from the trivialities of self so that one can be free to live in a large world. The decrease of ego allows one to apprehend the symmetry and beauty of the universe. In Buddhist thought, the historical quest of the individual to find meaning is equivalent to subsuming one's identity to the whole. Hence "I" is lost—yet also meaningfully found—within the interconnected sea of LIFE. Individuality coalesces into unity.

Buddhist thought is an attractive faith perspective for those seeking to grasp how holistic thinking illuminates our understanding of everyday life. A Western trekker to Nepal's Solu-Khumbu region, for example, cannot help but be impressed by how Buddhist precepts are manifest in traditional Sherpa culture. There is a fundamentally different quality to the way in which a typical Sherpa family relates to the environment in comparison with their Western counterparts. Respect for LIFE is a way of life.

Yet Buddhism holds within itself a paradox. If one can only apprehend totality by conquering the belief in one's separateness or individuality, how is it possible to point others to the pathway of illumination when they are still captive to the world of "self" and "other" (Ross, 1980: 20–21)? Stated alternatively, if the negation of oneself is a requirement for grasping the interconnectedness of the whole, how is the awareness of other selves possible? Apart from such awareness, passivity reigns the day, and the pathway to illumination lies beyond the reach of humanity.

To Western eyes the Buddhist paradox calls into question the precept that totality and individuality are trapped in a perpetual state of opposition. Why must the stature of the particular be diminished before the whole can be understood? Is it really possible to comprehend the whole without recognizing the way in which individuality shapes that conception of the whole?

The dilemma between individuality and totality—which finds heightened and eloquent expression in Buddhism—is a spiritual dilemma in the most profound sense of the term. It is also a dilemma for theoretical consciousness: How do we understand the significance of the individual in relation to the whole? People of good will who take seriously their responsibilities to nurture and protect the environment cannot avoid this question.

CONSTRUCTIVE INSIGNIFICANCE

The age-old question of "Who am I?" contains the seeds of a far more disturbing one: "What does it matter?" Gnawing doubts about one's own significance lie near the surface of most of what we think and do. Indeed, much of human culture and ideology is devoted to the artful denial of our own insignificance. When this question is forced upon us in one way or another, most of us resort to convenient off-the-shelf answers that offer a response to the question before the question is really asked.

In contemplating our individual roles within the vastly complex and ancient organism that we call the biosphere, it is difficult to see how any one individual or collectivity can be authentically significant within Earth's "lifescapes." The biological time between our age and the time when dinosaurs ruled the Earth is a brief interlude. The earliest known fossils, remarkably similar to blue-green algae, attest to the presence of life on earth 3.5 billion years ago. If we were to convert Earth time to a human calendar year, assigning the beginnings of life to January 1, the extinction of the dinosaurs took place on Christmas morning, December 25, at 5:20 A.M. The earliest humans came on the scene at 2:00 P.M. on December 31 and discovered agriculture a mere one and one-half minutes before midnight! The celebrated invention of writing took place 45 seconds later, and the Industrial Revolution began only 2 seconds before midnight. Our lives begin and end within a little more than one-half second of biological time. It is only in the disruption of ecosystems and the destruction of LIFE that humans can become significant in the ecological sense of the term. Such notoriety, however, appropriately does not satisfy the human yearning for significance.

Is it possible to embrace one's own insignificance while remaining passionately committed to purposeful living and the value of LIFE? For most of us, the question introduces a contradiction. How is it possible to live with purpose while admitting that insigificance characterizes the human condition? Must we persist in the denial of our own insignificance in order to thrive?

The solution to this predicament, I suggest, lies in redefining the way we understand significance. This is a difficult task for all of us, as we have grown used to the axioms of anthropocentrism in understanding ourselves in relation to the world. Yet, when one thinks about it, why should individual significance be a precondition for purposeful living in the first place? For example, most of the activities which busy human beings in modern societies engage in have little or no sense of purpose—especially when viewed from the clarifying perspective of the deathbed. That which really matters is often obscured by the tyranny of the immediate. Most people have to undergo some type of personal crisis before recognizing this truth.

What is significant about life is LIFE itself.[2] The miracle of life contains within itself a profound sense of purpose . . . a sense of purpose that becomes plainly evident when we contemplate the gift of life . . . a sense of purpose that inspires the passion for LIFE. The more we know about the natural world and interact with it, the more we grasp the wonder of our own life and the LIFE that surrounds us. To live life authentically is to respect LIFE unreservedly. A full life is life lived in the present tense; it savors each moment of existence. Even those who suffer from unremitting pain have the capacity to cherish the miracle of life—perhaps even more so.

What is striking about human civilization in our day is how much of our existence is life denying. We are constantly in the business of subordinating individual life and biospheric LIFE to a means-end calculus that orders the universe by human pride and prerogative. Such life denial can be sustained only by positioning ourselves over LIFE, exempting ourselves from the very processes that originally made life possible. How else could one explain the destruction of a rain forest?

What the human community needs most in our day is a perspective on life that allows us to grapple constructively with our own insignificance. I call this "constructive insignificance." By this I mean that humans must apprehend the wonder of LIFE and live constructively in response to that wonder—provoked into humility by their insignificance and energized by their passionate engagement with LIFE. As such, constructive insignificance is neither a drab nor a dismal outlook on life. Instead, it embraces the joy and purpose associated with the experience of life itself.

To a great extent the perspective of constructive insignificance resolves the dilemma posed by the Buddhist paradox. The whole is not apprehended by diminishing the stature of the individual or by progressively annihilating one's individuality. Instead, the whole is apprehended by emphasizing that which the part shares with the whole—in this case, the passion for LIFE. Nature's passion for life is articulated in countless ways with exquisite detail and overwhelming creativity. Inasmuch as we grasp the truth of the miraculous gift of life, we are able to affirm the joy of our individuality within the splendid matrix of LIFE that envelops us. Our captivation with bisopheric LIFE helps us reclaim the wonder of our own lives. Affirming the value of the whole in no way diminishes the part; it enlivens and strengthens the part.

As humans, our significance derives not from redefining the world on our own terms—telescoping biological time into human time and remaking habitats according to our own liking. Instead, our lives take on significance because we admit, in humility, to our insignificance and, in so doing, rediscover that which makes life authentically significant. In losing a conception of life, we find life.

AWE AND SACRED SPACE

A midnight walk along the ocean, a glaciated mountain glistening in the midday sun, the cacophony of voices that prevail across a tropical rain forest, and the splendor of a coral reef graced by multicolored denizens—these inspire awe within the human heart. Awe is a complex human emotion. We experience it when we encounter something beyond ourselves that engenders respect, fear, or wonder.

Awe has a certain reverential quality about it. To be in awe of a person is to recognize accomplishment or character that seems beyond our capabilities—provoking admiration and, sometimes, fear. To be in awe of some aspect of the natural world is to partake of its wonder. The more we know, the more enchanted we become. Who can study biology without pondering the magnificence of DNA—the universal tongue of all our planet's life forms? Who can learn of the bat's exquisite sonar guidance system without responding in amazement? Who can understand the marvelous precision of the human eye without cherishing the miracle of sight? Who can study the multitiered habitats of a tropical rain forest without feeling overwhelmed by the eloquent coexistence of mind-numbing diversity and interdependence?

The awe we feel in response to the magnificence of the natural world is similar to the awe experienced by the faithful when entering a medieval cathedral. Take, for example, a rain forest. As biologist E. O. Wilson puts it, "Seen on foot, a rain forest is like the nave of a cathedral, a thing of reverential beauty yet with much of its splendor out of reach in the towers and illuminated clerestories high above" (1991b: 80). This cathedral, like those of human creation, reverberates with sounds that provoke reverence. Like Johann Sebastian Bach's remarkable chorale prelude for the Lord's Prayer, the music of the rain forest is discordant to the ear because of its overwhelming complexity. Yet it displays an underlying coherence for those who have ears to hear (Diamond, 1990).

Not only contemporary biologists, but also the indigenous peoples of Amazonia understand the sacred spaces of the forest. The spirits of the forest function as the equivalent of supernatural game wardens, placing segments of the forest off limits to hunting and fishing, as well as encouraging methods of harvesting that conserve nature. Consider, for example, the role of *mãe de seringa*—the "mother of rubber trees." She is a short woman with long hair who makes surprise appearances to rubber tappers on forest trails—admonishing them, when necessary, not to slice the rubber trees too deeply. The following story is told along the upper Rio Negro in Brazil about one rubber tapper who became overzealous in collecting sap from the trees.

One day, he was deeply gouging a rubber tree when something tapped him on the shoulder. He spun around and came face to face with an ugly, light-skinned woman with fair, tousled hair. "You are mistreating my daughters, draining away all their blood. Stop cutting them!" commanded the hag.

The surprised tapper explained his predicament. "My sponsor is pressuring me to come up with more rubber to pay off my debts, but the trees seem to be drying up and I am broke." Reacting with compassion, *mãe de seringa* offered to leave a large ball of smoked latex in the man's fire-hut each night from October until March if he would leave her daughters alone. She warned him never to tell anyone about their accord, however, for if he did, he would join her. The tapper thus spent the remainder of the season loafing, hunting, fishing, and making easy money.

During the next tapping cycle, the secret bargain was renewed.

When the tapper brought his first load of rubber to his patron's floating store, the boss was suitably impressed. His credit thus restored, the tapper immediately bought a bottle of cachaça, a potent alcoholic drink made from sugar cane. The crystal-clear liquid soon took effect and the effusive tapper began boasting about his special relationship with the mother of rubber trees. Shortly after returning to his makeshift hut, the indiscreet tapper ventured into the forest where he was bitten by a venomous snake. A neighbor later found his lifeless body slumped on a trail leading to some rubber trees. (Smith, 1983)

Some concept of sacred space can be found in every culture. It is a way for us to give awe and transcendence a geographic locale. Sacred space, in this respect, becomes a kind of geographic crossroads between the individual and the vast totality that remains outside his or her grasp. There is even some evidence to suggest that some traditionally sacred places, such as the springs at Lourdes or the sacred mountains of the Incas, may stimulate visions because of anomalous electromagnetic fields associated with the geology of the area (Gallagher, 1993).

Whether sacred space is conceived as a place imbued with ritual significance or as a place set apart for some special purpose, the creation of sacred space affords the opportunity for humans to experience awe and to touch that which remains transcendent. Visitors, for example, to the Buddhist monastery at Tengboche, Nepal (12,687 feet) cannot help but be impressed by the drama created by the nearby mountain, Ama Dablam, which dominates the immediate landscape at 22,510 feet. No one is permitted to climb Ama Dablam. It is a sacred mountain. The Buddhist monk who greets the beginning of a new day by striking an immense gong in the brisk mountain air of dawn looks out from the top of the monastery toward Ama Dablam. The sacredness of the mountain imbues this daily ritual with an unforgettable quality, surpassing that of the most breathtaking medieval cathedral.

For modern societies, the most important manifestation of sacred space, from an environmental standpoint, is the institution of the national park. A concept born on American soil, the national park ideal is perhaps our most salutary export product, next to that of constitutional democracy. It is an ideal, however, that must be contextualized and redefined to reflect the needs of local cultures and the rights of indigenous peoples (Harmon, 1987). In particular, the concept of mixed-use national parks or integrated conservation-development projects, which give indigenous peoples an incentive to preserve the environment, is especially promising.[3]

Take, for example, Zimbabwe's Matobo National Park. The lush vegetation of the park stands in sharp contrast to the overgrazed and densely populated areas immediately adjacent to the park boundaries. Without the cooperation of the pastoralists who encircle the park (and once freely traversed the park with their cattle herds), it would be quite impossible to maintain the park as a viable ecological preserve. Recognizing the need for such cooperation, the park authorities in 1962 met with local villages and negotiated a trade. In exchange for their cooperation in protecting the park against poaching, cattle herding, and the setting of fires, villagers were given access to an abundant renewal resource within the park that was in short supply: thatch. As the primary roofing material for the region, thatch was an important economic resource that was becoming increasingly scarce as a result of overgrazing. Each village was licensed to cut a predetermined quota of thatch, paying back the park one bundle for every ten collected. This policy has generated an income of between $20,000 and $60,000 per year to the surrounding villages and has effectively discouraged poaching and wild fires within the park (McNeely, 1988: 156). Zimbabwe, and its northern neighbor Zambia, are currently experimenting with similar village-based programs for wildlife management—in effect, empowering local villages to be responsible for protecting wildlife while receiving a portion of the economic benefit (e.g., safari hunting fees). As a consequence, the incidence of poaching has decreased dramatically (Worldwatch Institute, 1992: 22).

Another example of forging creative links between national parks and local communities comes from the oldest national park in Thailand: Khao Yai. The village of Ban Sap Tai, adjacent to the park, had developed a reputation for poaching wildlife within the park's boundaries, as well as cultivating park lands for their own benefit. In 1985 a pilot program was initiated to address this problem. The project initially was focused on encouraging conservation of the park's resources by demonstrating the economic benefits of "nature tourism" for the villagers. For example, villagers who assisted as guides or porters on treks within the park received about three times the average wage for the region. This trekking program was later broadened to include the establishment of a village-based Environmental Protection Society, funded by Agro Action, a German nongovernmental organization. The Environmental Protection Society is a distinctively hybrid cooperative that combines the functions of being part credit union, part educational center, and part business enterprise. All villagers who own land in the community can belong to the cooperative, provided that they pledge to uphold the park's regulations. The success of the cooperative has been remarkable. From 1985 to 1987, the membership within the cooperative nearly tripled, representing over 70 percent of the villagers. Poaching has been dramatically reduced, and farms that were formerly inside the park's boundaries have been vacated (McNeely, 1988: 153–55).

The search for sacred spaces by Western tourists who have grown weary of colorless concrete and tedious tourist traps has generated a small yet promising industry that has great potential to encourage conservation—if it is done right. For example, it was estimated that in 1988 "ecotourism" to Third World destinations was a $12 billion industry. Approximately one-half of these revenues were captured within their local economies (World Resources Institute, 1992: 140). Ecotourism is now a major source of foreign exchange for such countries as Ecuador and Kenya. In 1993 Ecuador alone expects to attract 30,000 tourists from North America and Europe to its rain forests. Altogether ecotourism within Amazonia is increasing at a rate of 12 percent per year (Brooke, 1993). The celebrated Monteverde Cloud Forest Reserve of Costa Rica alone earned more than $1 million in profits from ecotourism. A women's cooperative that markets local crafts to Monteverde's tourists has annual revenues in excess of $50,000 (Stammer, 1992). In Nepal ecotourism has been primarily responsible for saving the great Indian rhinoceros through the establishment of the Royal Chitwan National Park. One ecotourist business that operates in the park, the Tiger Mountain Group, has even established a trust fund designed to promote conservation education in the villages that surround the park (McNeely, 1988: 92, 118).

At the village level, the benefits of ecotourism can be a powerful impetus for ecological preservation. Take, for example, two villages on the island of Tavenui in Fiji. Both villages wanted to raise money to pay for their children's education and to upgrade their homes. Both villages had legal claim to large tracts of the surrounding rain forest. The first village elected to sell its inheritance of the rain forest to loggers for a substantial, yet short-lived, profit. The second village decided to utilize its segment of the forest—including a lake and beautiful waterfalls—for nature tourism. During the first nine months, the second village generated $8,000 for their project with the promise of an ecological annuity for many years to come (Stammer, 1992).

Among the more creative expressions of the search for sacred space are attempts by urban planners to replicate living ecological zones within the architectural confines of concrete and glass. From small neighborhood parks to communally managed urban gardens to innovative rooftop gardens, there is incredible potential for densely populated urban areas to be remade into biologically inviting habitats. Take, for example, the well-established institution of the courtyard in densely populated traditional settlements. The courtyard provides an intimate gathering place for conversation and reflection, appointed with greenery and protected from the noise and dust of the outside world. Moreover, the courtyard takes up much less space than the modern backyard and is a more effective structure for communal interaction (Eibl-Eibesfeldt, 1989: 645). The contemporary atriums in plush hotels, office buildings, and shopping malls attest to our desire for courtyard-like spaces. Nevertheless, the cost-effective institution of the courtyard has rarely been utilized in modern housing developments.

Rooftop gardens, parks, and greenhouses that utilize solid-state hydroponics also hold tremendous promise for recreating our cities. Over the past several years, the Gaia Institute in New York City has perfected a lightweight soil for rooftop gardens made primarily from finely ground recycled styrofoam. Whereas a cubic foot of ordinary soil weighs approximately 100 pounds, the Gaia Institute substitute weighs only from 9 to 22 pounds per cubic foot (Walter, 1992). The potential of such rooftop agriculture for economic and psychological well-being is immense. It is only beginning to be explored. For example, in Oakland, California, there is a park complete with grass and trees built on the top of a six-story parking garage (Register, 1992).

TOWARD ECOLOGICAL LITERACY

To experience sacred space is to give awe a foothold in our lives, to affirm the integral connection between ourselves and all that is beyond us, and to receive a glimpse of what life can be without Berlin Walls. Berlin Walls rob us of our passion for life. Wall-building projects that condemn humans to an ecological version of solitary confinement destroy the human spirit. To realize one's humanity at the end of the twentieth century is to grapple with the redefinition of our place in the world—and, in so doing, to rediscover our passion for life.

In many respects, the redefinition that is taking place in our day is as fundamental to human consciousness and culture as was the invention of writing some 5,000 years ago. The cuneiform tablets inscribed within the city gates of Uruk, Mesopotamia, began a revolution that would redefine how human communities saw themselves and the world around them for millenia. The empowerment associated with the simple acts of reading and writing changed the face of humanity forever. Ecological literacy holds a similar potential for humanity, particularly when linked to contemporary information technologies (Annis, 1992).

Just as reading and writing empowered the earliest cuneiform scribes, so also the new ecological literacy empowers people to live life with fullness and intensity. Those who discover the power of literacy never renounce it. Whereas the ability to read and write was the entitlement of only a privileged few until the past 200 years or so, the new ecological literacy, and the empowerment it brings, must not be captive to elites but open to all. It is only then that the new literacy will have the power to reconfigure our relationship to the environment.

Ironically we have much to learn about the new literacy from the very groups that we all too readily deem backward and illiterate. Many indigenous peoples have a keen sense of their moral obligation to preserve the forest and its resources for future generations (World Bank, 1992: 94). A saying is commonly heard among the indigenous peoples of Colombia that ''the difference between a colonist [a nonIndian settler] and an Indian is that the colonist wants to leave money for his children and that the Indians want to leave forests for their children.''

A recent survey conducted by a National Geographic Society team in Central America, conducted in 1992, examined the relationship between Indian lands and forest cover. The team confirmed the observation of a Kuna Indian from Panama: ''Where there are forests there are indigenous peoples, and where there are indigenous people there are forests.'' The Philippines and Thailand provide ample evidence of this relationship. Only about one-third of what the government has offically designated as forest are currently forest covered. Those forests that are still intact have been preserved by tribal peoples (Worldwatch Institute, 1993: 85). A similar brand of ecological sensitivity is also evidenced—to varying degrees—among low-income populations. It appears that the vulnerability caused by poverty, in the right circumstances, has the effect of mobilizing otherwise fragmented and apathetic populations for meaningful environmental advocacy (Broad, 1994).

In light of the above, there is a need for strong movements of indigenous peoples and other vulnerable groups which have the capability to make their case effectively to the central government. For example, as a result of effective demonstrations and lobbying by Ecuador's Indian federations, the government granted legal title to over three million acres of forest homeland to its indigenous peoples. The land titles are given to the indigenous communities themselves, not to individuals. Moreover, the land cannot be sold under any circumstances. Virtually all development is banned, aside from oil exploration, without the approval of the communities themselves (Farah, 1992; Worldwatch Institute, 1993: 97–98).

Upholding the rights of indigenous peoples translates into protecting the Earth's biodiversity. Fortunately, many of the dozen or so countries that qualify as "megadiversity countries" (collectively accounting for between 60 and 70 percent of the Earth's total biodiversity) are also characterized by a high level of cultural diversity among indigenous peoples. For example, approximately 670 languages are spoken in the nation of Indonesia. As a megadiversity country, there are also an estimated 22,900 species of mammals, birds, amphibians, reptiles, and flowering plants within its borders. A similar correlation between human diversity and species diversity can be found in India, Australia, Mexico, Zaire, and Brazil (McNeeley et al., 1990: 88–89; Worldwatch Institute, 1993: 86).

One of the distinguishing characteristics of traditional societies in relation to their modern counterparts is their commitment to the land. As the World Council of Indigenous Peoples declared, "Next to shooting Indigenous Peoples, the surest way to kill us is to separate us from our part of the Earth" (Worldwatch Institute, 1993: 85). By contrast, modern industrialized societies embody a form of cultural homelessness, fostered by the convergence of widespread geographic mobility and enthusiasm for reengineering habitats. This lack of commitment to the land conspires with our penchant to objectify nature to produce a form of alienation far more profound than the sort described by Karl Marx in relation to industrial society. We are rarely fully present at home, because home is typically a place without enduring connections—apart from being geographically close to one's family, decorating one's habitat with souvenirs from the past, or owning a residence that has land. While most citizens of modern industrial societies can never fully experience home in the sense of an Iowan farmer or a Sherpa yak herder, we can overcome our sense of homelessness by making substantive investments in our local habitats. I will have more to say about this momentarily in connection with the concept of Earth communities.

The degree to which ecological literacy has the power to transform us will be directly related to our ability to remove incentives for illiteracy.

At present a large amount of our economic resources is devoted to subsidizing activities that are harmful to the environment (World Resources Institute, 1994: 23–24). Whereas most people despise parasitic free riders when it comes to the workplace, the welfare system, or the defense industry, we devote little or no attention to the problem of environmental free riders. There are a multitude of ways, for example, that public policy creates hidden subsidies for environmental devastation. Economic policy, tax policy, energy policy, transportation policy, and policies regarding housing and zoning all contribute to the huge subsidy that we pay to encourage ecological devastation. As a result of these subsidies, the adage that "good ethics" is "good business" is usually not true when it comes to making business decisions in relation to the environment (Hoffman, 1991). Much of the new ecological literacy must target these costly subsidies, raising public consciousness and creating a political climate conducive to change.

For ecological literacy to take hold, education is the key. No group should be excluded. It is just as important for children to be literate as adults. Moreover, as with traditional literacy programs, we must find creative ways to reach out to assist adults in overcoming their ecological illiteracy through nonformal educational programs.

Certainly the requirement of ecology courses at the primary, junior high, and high school levels—along with an overall upgrading of science curricula—would be an important way to promote the new literacy. Significantly, the unrestrained enthusiasm of children for environmental education—and their impact on parental attitudes—has surprised more than a few elementary school teachers in the United States. Similarly, plant-a-tree programs for school children in developing countries have been extremely successful. When near "virtual reality" technology becomes available in school systems, it will establish a new standard for environmental education. If children are able to experience what it is like to fly through a rain forest or navigate a coral reef, indentifying species along the way, they will, as adults, have some grasp of what is at stake when a rain forest or coral reef is destroyed.

At the college level, there are a number of imaginative opportunities to promote ecological literacy. For example, the creation of required college-level courses on Integrative Ecology (combining ecology with the humanities and social sciences) could become an exciting component of a student's core curriculum. Furthermore, there is tremendous potential for academic institutions to work directly with the national park system in creating "green classrooms" or "classrooms with trees" for both traditional college students and nontraditional continuing education. Also, semester-abroad experiences for college students, involving a study experience in a rain forest or some exotic habitat, hold tremendous potential for ecological education. One could likewise envision the development of international internship programs that would offer scholarships to university students from developing countries to work at the side of professionals experienced in wildlife and park management.

In terms of adult education, the media have a special role to play in promoting ecological literacy. In particular, there is a need to spotlight habitat degradation with the same passion that would be devoted to conducting investigative reporting on a corrupt politician or a defective product. Take the problem of biodiversity. There is little understanding among the general public about what is at stake. What would happen, though, if the death of a rare plant or animal species were reported in the media with the same frequency as the death of significant personalities on the evening news? At first, this might seem odd to the audiences of Tom Brokaw or Peter Jennings, but after a while such reports would arouse people's curiosity and begin to penetrate their consciousness, particularly if reports about the death of one species were juxtaposed by a report about the promise of another species for medical research. In response to the rejoinder "who cares?" or "the media tell the public what it wants to hear," it should be emphasized that the media have traditionally felt a responsibility to position themselves at the frontier of scientific research and knowledge, exposing the public to the wonders of the technological age. During the early years of the space program, for example, Walter Cronkite's broadcasts from Cape Canaveral dominated American televisions for hours at a time. The prospect of exploring our moon and other planets demanded media that were willing to play the role of educator—not simply entertainer. The challenge to preserve our Earth's biosphere demands nothing less.

One initiative that would help focus media attention around the environment would be the establishment of a Nobel Prize for Environmental Responsibility—the logical extension of the Nobel Peace Prize in the ecological sphere. Such an award would provide immense moral support to the work of environmentalists, ecologists, biologists, policy makers, and others who are actively involved in making peace with the biosphere. Many unsung heroes and heroines work in virtual obscurity for environmental justice. The work of Kenya's Wangari Maathai, Malaysia's An-

derson Mutang, and India's Medha Patkar deserves the sort of
international recognition that a Nobel Prize for Environmental Responsibility would bring (Toufexis, 1992).

The business community also has an important role to play in edu-
cating the public about the environment. For example, one should not
underestimate the role of such companies as The Body Shop and Ben
and Jerry's in promoting environmental awareness. Both companies have
a well-publicized commitment to environmental responsibility. Moreover, both firms have been remarkably prosperous in the marketplace,
thereby demonstrating that environmental responsibility *can* be recon-
ciled with entrepreneurial success (Worldwatch Institute, 1993: 185–86).

Similarly, the creative partnership between environmental organizations such as the Environmental Defense Fund, and corporate giants,
such as McDonald's, is very encouraging. Instead of remaining adver-
saries, the Environmental Defense Fund and McDonald's decided to
team up to work on ways to make fast food more ecologically friendly.
Through the partnership, McDonald's benefited from a significantly improved environmental image in the public's mind. More important, the
Environmental Defense Fund profited by the opportunity to promote ecological awareness through the giant fast-food chain, encouraging
McDonald's suppliers and its staff and customers to consider how their
behavior impacts the environment (*The Economist*, 1992b).

Innovative alliances between environmental organizations and business corporations should be strongly encouraged by government and the
foundation community. One promising stimulus in this regard is the
growing phenomenon of environmental labeling within European countries, Japan, Canada, and the United States. By providing consumers with
information that helps them make connections between environmental
protection and what they buy, authentic environmental labeling gives
people the satisfaction of making a small contribution to ecological well-
being through their purchasing decisions. Of greater consequence, environmental labeling initiatives provide a market-based incentive for
companies to exercise more environmental responsibility (Organization
for Economic Cooperation and Development, 1991; Worldwatch Institute, 1992: 184).

The spread of ecological literacy through formal and nonformal educational initiatives is a crucial ingredient in confronting the ecological
challenges of the twenty-first century. Yet, by itself, ecological literacy
is not enough. The economic, social, and political change that Earth
requires of humanity is simply too much too soon for education to handle

the burden alone. Education takes time, and time is something we do not have when it comes to preserving the rain forests and coral reefs.

THINKING STRATEGICALLY ABOUT ENVIRONMENTAL REFORM

The promotion of environmental change on a global scale requires a multidimensional strategy. Such a strategy would encourage simultaneous and mutually reinforcing change at both the micro-level and the macro-level. Most important, such a strategy would have the capacity to spark "cascade" or "wildfire" effects, where seemingly insignificant change is amplified and spread in such a way that it "snowballs" into substantial change (Lewis, 1969: 88). This is similar to the process of autocatalysis in chemistry, where a by-product of a chemical reaction functions as a catalyst for the reaction as a whole, thereby accelerating the pace of the reaction. What would be the general outlines of such a global strategy?

Two of my colleagues, Elizabeth Morgan and Grant Power, and I recently completed a study on existing proposals for global institutional change, culled from a wide range of literatures, including development studies, political science, economics, environmental studies, social ethics, and international law (Morgan, Power, and Weigel, 1993). On the basis of this survey, we constructed a heuristic typology of what we call "strategic pathways" for global institutional reform. These pathways serve as general categories for specific "transition strategies," which we define as programs of action seeking to move the world from a specified present state of affairs to a preferred future position. We identified four criteria for a complete transition strategy, recognizing that not all of our examples would satisfy all four. These criteria are as follows:

1. A problematic present condition (Point A)
2. A desirable future condition (Point B)
3. The overarching process, trajectory, or direction in which relevant actors could move from Point A to Point B
4. A set of identifiable steps, events, or components that are indispensable to the effective implementation of the strategy

The literature that we surveyed contained a range of transition strategies which we were able to classify within six general categories:

1. Strategies that focused on the creation of global issue regimes
2. Strategies that focused on more developed countries (MDCs) as initiators of change
3. Strategies that focused on less developed countries (LDCs) as initiators of change
4. Strategies that focused on the creation of alternative institutions
5. Strategies that focused on the resolution of international conflict
6. Strategies that focused on global value transformation

Without going into the details of the study, we concluded that strategic thinking on global change needed to take advantage of linkages among various strategies in order to capitalize on their respective strengths, as well as their natural affinity to either micro- or macro-level policy environments. Instead of being wedded to a particular strategy for change, a bundle of mixed strategies created new options that were unavailable when the strategies are taken separately. In thinking about the sort of institutional change that the Prime Directive would require across the globe, there is no question that it will be necessary to pursue a multidimensional strategy that draws from the best of each strategic pathway.

Four strategic pathways, in particular, seem well suited for the long and arduous task of shaping both national and international institutions in accordance with the Prime Directive. Two of these pathways are oriented primarily around change at the micro-level (e.g., individual lifestyles, community-based movements, nongovernmental organizations). They are the "global value transformation" pathway and the "creation of alternative institutions" pathway. The other two strategic pathways are focused on change at the macro-level (e.g., domestic policy, foreign policy, intergovernmental organizations). They are the "global issue regimes" pathway and the "more developed countries as initiators of change" pathway. Let us look briefly at how a combination of these strategic pathways could initiate cascading institutional change.

PROMOTING CASCADING CHANGE AT THE MICRO-LEVEL

One of the primary problems associated with institutional change, particularly in a global context, is that people are generally overcome with feelings of powerlessness when they try to effect meaningful change at the macro-level. It stands to reason that one's enthusiasm for becoming invested in social change dwindles in proportion to the perceived complexity and intransigence of the envisioned change. The more one learns about a particular problem, the more one appreciates how little one person can do. This, in turn, takes its toll on concerned persons with vision, robbing them of their "can do" enthusiasm and replacing it with apathy and skepticism. Consequently, one of the requirements of any robust multidimensional strategy is that concerned persons must be convinced that they, as individuals, have an opportunity to make an authentic and concrete contribution to secure meaningful change.

One of our strategic pathways that is especially amenable to bringing change down to the level of individuals and small groups is the "global value transformation" pathway. At the heart of the global value transformation pathway lies a belief in the power of ideas to influence the conduct not only of individuals and families, but also nations and international organizations—mobilizing untapped resources of imagination and energy. Hence this pathway for global reform is well suited to bold, visionary proposals for change that depend primarily on a change of individual and social values. Such value-oriented change could be promoted in a one-on-one context through conversation or education, or it may occur within the less desirable yet more forceful context of the lessons we learn through environmental catastrophe (Worldwatch, 1992: 177).

Of course, detractors from this point of view are often quick to dismiss adherents of this pathway as idealists or utopian thinkers. Nonetheless, the power of ideas should not be underestimated as key elements in a reform package. After all, the massive sociopolitical movements of human history can be attributed, in part, to the enticement of ideals (e.g., the abolition of institutionalized slavery, the nearly universal rejection of racism as a political/economic organizing principle, the breakup of colonial empires, the rise and fall of state communism). Unfortunately, though, our understanding of the transmission of ideas and their impact on actual behavior is in its infancy (Sjoberg, 1989). Changed values do not necessarily result in changed behavior—a fact often conveniently overlooked by proponents of this value-based strategy for change. This problem is intensified when the emphasis on value change clouds one's assessment of the fact that well-entrenched sources of economic and political power have strong vested interests to preserve the status quo.

A notable example of a values-oriented strategy to environmental reform is the Washington-based Worldwatch Institute. Its highly influential *State of the World* reports portray in vivid terms the economic and ecological threats posed by both the consumption sickness of advanced industrial societies and the unsustainable modes of development being practiced by the Third World. The discussion ranges from the demographic and geographic problems of overpopulation, deforestation, and desertification to the social and political issues of pointless militarism, addictive consumerism, and political apathy. Throughout the pages of its *State of the World* reports, the Worldwatch Institute advocates the idea of a ''sustainable society,'' a society that ''satisfies its needs without jeopardizing the prospects of future generations'' (Worldwatch Institute, 1990: 173). Such a society would be characterized by the limited use of fossil fuels and a decreasing reliance on nuclear energy; the increased utilization of solar and wind energy, extensive recycling, and agroforestry in developing countries; and the diffusion of information-age home-based workplaces. Worldwatch's strategy relies strongly on the way in which environmentally sound individual and social values can produce a sustainable global economy. The achievement of Worldwatch's sustainable economy on a global scale requires nothing less than a wholesale reorientation of individual and social values. This reorientation calls for us to abandon our habits of consumptive development and to embrace the vision of a world that is less harmful to the planet and more satisfying to inhabit. ''The goal of the cold war was to get others to change their values and behavior, but winning the battle to save the planet depends on changing our own values and behavior.'' (Worldwatch Institute, 1991: 4).

One of the strategic problems associated with a values-based approach to environmental change is the need to embody one's values in tangible actions beyond the purchase of environmentally friendly products and recycling. Those who embrace a new set of environmental values need concrete and constructive ways to channel and reinforce the values they have embraced. Otherwise environmental responsibility will become captive to the whims of passing fads.

A second strategic pathway within our typology—the creation of alternative institutions pathway—helps remedy this problem. This stategic pathway is focused on the potential of alternative institutions, generally nongovernmental institutions, to help reorder the priorities of dominant economic, social, and political institutions. In effect, the reform process is understood from the standpoint of a grassroots or people's movement ethos, where the creation of alternative institutions largely bypasses (and occasionally usurps) traditional governmental prerogatives and domains—ultimately holding the potential to co-opt or redirect the dominant institutions of society, including governmental institutions.

Perhaps the most prominent manifestation of this strategic pathway is the modern nongovernmental organization (NGO) and its success in facilitating authentic participation in the process of development. For instance, David Korten (1990) makes a persuasive argument for the power of grassroots organizations to impact public policy at the national and international levels. He envisions a four-stage growth process in the NGO community that moves from a first-generation emphasis on relief and welfare to a second-generation concern for "empowerment strategies" that promote small-scale self-reliant local development. From this platform, Korten envisions a third generation for the NGO community focused on efforts to change local, national, and international policies in the interest of promoting sustainable development. Finally, with the eventual development of functional alliances between northern and southern NGOs and among NGOs that share common policy interests, Korten believes that we will see a fourth generation of NGOs that will usher in a "people-centered" approach to development on a global scale.

The recent work of Julie Fisher (1993), which significantly extended Korten's work, has focused the discussion around the challenge of sustainable development. Fisher examines the close interrelationships between local grassroots organizations (GROs) and grassroots support organizations (GRSOs). In a forthcoming companion volume, she will focus on the relationship of NGOs to governments and outside donors, examining the ability of NGOs to facilitate sustainable development. Altogether Fisher estimates that there are between 30,000 and 35,000 active grassroots organizations in the Third World. In São Paulo, Brazil, alone, there are over 1,300 neighborhood improvement associations active in planting community gardens, recycling garbage, constructing water taps, and building street gutters (1993: 30). In the Majarashtra and Gujarat states of India, the Swadhyaya movement is active in hundreds of villages. In Sanskrit, the word *Swadhyaya* means the study of self. The movement emphasizes the importance of self-reliance. Its projects include the establishment of small farms that devote their profits to the purchase of agricultural implements for the entire community. Moreover, believing that trees testify to the presence of God, adherents of the Swadhyaya movement routinely conduct tree planting rituals, thereby practicing environmental responsibility as a natural outcome of their spirituality (69).

Similar grassroots movements around the world are trying to address what environmental responsibility means to their local communities. For example, in Burkina Faso, the village-based Naam movement is making significant inroads in ecodevelopment across the country. Traditionally the term *naam* referred to the sharing of tasks by young people for the preparation of feasts and other communal tasks. Later a Burkina Faso sociologist revived the tradition and applied it more generally to involve the entire village in a variety of community projects, including well construction, the maintenance of communal forests, and land-shaping projects to control soil erosion. Since its beginnings, the movement has seen remarkable growth. In 1973 there were only about 100 active Naam groups, but by 1987, that number had grown to more than 2,500 groups. Similarly, the Green Belt Movement in Kenya promotes environmental protection by encouraging women to plant tree seedlings in public areas for the purpose of forming tree belts. Established by Wangari Maathai, a Kenyan environmentalist and women's rights leader, the Green Belt Movement has enlisted the help of some 50,000 Kenyan women and has planted some 10 million trees in 1,000 tree belts (World Resources Institute, 1992: 224–26).

Within industrialized countries, NGOs play a critical role in mobilizing public opinion and disseminating information to the public about ecological concerns. Moreover, NGOs provide a number of helpful avenues for people to become active in preserving the environment (e.g., urban gardening programs, scouting, recycling programs, car pooling, adopt-a-highway initiatives). There is a considerable need, however, to broaden the scope of programs and initiatives designed to foster environmental awareness within industrialized countries.

One promising alternative available to both industrialized and less developed countries would be to experiment with the notion of Earth communities. An Earth community refers to any group—a family, school, church, town, city, or state—that is actively seeking to accomplish two objectives: (1) to establish meaningful connections between people and natural habitats within and around the community and (2) to promote respect for the Earth's ecology outside the "boundaries" of the community. Therefore, each Earth community has an inward-looking objective and an outward-looking objective. Virtually any existing group, organization, or locale could designate itself an Earth community if it made a concerted attempt to work on these objectives. A Girl Scout troop, a high school, a church or synagogue, a small town, or a large city could become an Earth community. It would be especially exciting if such Earth communities, as part of their second objective, could develop sister Earth communities in other countries—similar to more familiar "sister city" initiatives. At the very least, Earth communities could promote socially responsible investing in so-called green investments. Mutual funds, such as the Calvert Group mutual funds and the Global Environment Fund in the United States or Britain's Merlin Ecology Fund, provide individual investors with the opportunity to gain environmental objectives through their investments (Worldwatch Institute, 1993: 196).

The potential for the concept of Earth communities to catch on is enhanced by the fact that nearly every contemporary issue facing modern civilization has an underlying environmental component. For example, the issues of peace and justice, human rights, women's issues, urban planning, poverty, and economic development all touch on the issue of ecological responsibility in one way or another. There would be no inconsistency in redefining these concerns in such a way that their ecological dimension was more pronounced. Respecting the essential interdependence of the Earth's biosphere is not merely one more special interest—it is an overriding interest that impacts practically everything we do.

By joining the global values transformation strategic pathway with the creation of alternative institutions strategic pathway, we are able to reinforce and channel meaningful changes in individual and social values into tangible actions and policies. Without such concretization, talk of value transformation will amount to little more than words or merely symbolic acts of environmental concern. By contrast, the power unleashed by people creatively putting hands and feet to their ideas should never be underestimated. Moreover, the creation of alternative institutions in both developing countries and industrialized societies may have a highly beneficial "demonstration" or "modeling" effect for reforming governmentally based environmental intiatives.

PROMOTING CASCADING CHANGE AT THE MACRO-LEVEL

Seeking change at the macro-level is understandably considerably more difficult to do in relation to environmental responsibility, primarily because of the substantial complexity associated with global environmental regulation and the short-term benefits that are characteristically associated with noncooperation. Nonetheless one can see some significant openings for cascading change at the macro-level.

One strategic pathway for global institutional change that is
particularly relevant to environmental concerns is the concept of ''global
issue regimes'' (Coate, 1982). The function of such regimes is to institutionalize a set of norms, principles, or procedures for guiding the conduct of states and international organizations with respect to a particular
issue area (e.g., environmental degradation, ozone depletion, absolute
poverty, the debt crisis, resource exploitation in Antarctica). Beyond
specifying the norms that ought to guide global behavior on an issue,
these strategies usually propose the formation of a global institution that
embodies these norms and is vested with sufficient authority for their
implementation. On occasion, global issue regime theorists may also
specify a set of concrete steps in implementing a new issue regime.

A starting point for many advocates of global issue regimes is a conviction that the big issues of the late twentieth century lie well beyond
the traditional borders of sovereign states and require substantial interstate cooperation through the creation of state-sponsored international
regimes. Because current international institutions are either weakly empowered and marginalized by nation-states or are based on sets of norms
incapable of handling these international issues, some type of institutional reconfiguration is required. This could involve the creation of a
new intergovernmental organization, the remaking of an existing institution, or the adoption of new multilateral treaties with some sanctioning
machinery.

Global issue regime theorists predict that nation-states will be con-
senting parties (albeit perhaps grudgingly so) to this process of institutional reconfiguration, presumably as a means of safeguarding the
legitimacy of nation-states in the face of cascading crises within the
international order. Moreover, these theorists hold that such institutional
reconfiguration will help diffuse and institutionalize emergent norms in
the international community (e.g., the human rights concept, environmental protection) that appear to be indispensable ingredients in crafting
solutions to specific problems.

A recent example of an emergent global issue regime is the regulatory bodies and institutional mechanisms that were left in place as a
result of the 1992 Rio Earth Summit (see World Resources Institute,
1994: 224–26). These ongoing mechanisms ultimately may be of much
more significance to the world community than the agreements that
proceeded from the Rio Summit itself. One can hope that the formation of the UN Commission on Global Governance, and their proposed

1998 World Conference on Governance, bodes well for the future of global issue regimes.[4]

There is no doubt, however, that the most successful illustration of an environmentally focused global issue regime to date is that of the land-mark Montreal Protocol of 1987 and its subsequent amendments—calling for the elimination of chlorofluorocarbon (CFC) production by 1996. For those who regard multilateral international negotiations as being "too little too late" to be effective instruments for environmental regulation, the Montreal Protocol is living proof that the international community can respond to new challenges in a timely fashion. It was not until the discovery of the ozone hole in Antarctica in 1985 that the problem of ozone depletion due to CFC production drew widespread attention in the press. As a result of the 1987 protocol, the worldwide production of CFCs fell by 46 percent from 1988 to 1991 (Worldwatch Institute, 1993: 186–87). It is even more encouraging that recent atmospheric monitoring indicates a significant decrease in the growth rate of CFC-11 and CFC-12 (used for refrigeration and air conditioning, as well as the production of aerosols and foams). This trend would suggest that CFC concentrations in the upper atmosphere will reach "a maximum before the turn of the century, and then begin to decline" (Elkins et al., 1993). A key to the success of the Montreal Protocol was the creation of a multilateral fund by industrialized countries to assist developing countries in making the transition to CFC substitutes (World Resources Institute, 1992: 152; Worldwatch Institute, 1992: 161).[5]

In using global issue regimes to address international environmental problems, one suspects that there will be a strong need to rely upon the political leadership of the more developed countries for two reasons. First, the mass publics of industrialized countries have considerably more information available to them concerning global environmental concerns and are generally more influential in shaping government policy than their counterparts in developing countries. For example, public concern about ozone depletion within the United States seems to be have been a primary factor in motivating the U.S. government to encourage the initial negotiations that led to the Montreal Protocol (Worldwatch Institute, 1992: 161). Second, most global environmental problems are unevenly distributed geographically. For example, all of the so-called mega-diversity countries (i.e., countries with unusually high numbers of plant and animal species) are in the Third World; Australia is the only exception (McNeely et al., 1990: 88). For the richer countries of the world to encourage the poorer countries to protect their unique habitats will, no doubt, require some financial commitment on the part of the industrialized countries. This could take the form of direct financial assistance, similar to the Montreal Protocol's multilateral fund. For example, China has requested approximately $2.1 billion for the phaseout of CFCs and the production of higher cost substitutes for refrigeration (World Resources Institute, 1994: 78). The industrialized countries could also make creative use of "debt-for-nature" swaps, involving the use of a portion of outstanding Third World debt to leverage initiatives and policies that protect the environment (World Resources Institute, 1992: 123, 226–27, 309).

In light of the above considerations, it seems reasonable to link a global issue regimes strategy with the expectation that much of the needed political leadership will come from the industrialized world. In this regard, one should not underestimate both the positive and negative consequences of the so-called demonstration effect by industrialized countries for the rest of the world. A lack of interest in ecological preservation by industrialized countries, for example, will severely weaken their efforts to encourage conservation programs in developing countries. Alternatively, environmental progress in the First World (e.g., the use of unleaded gasoline, air quality monitoring, toxic waste disposal) can have a positive impact on environmental standards in the Third World. One should not underestimate the beneficial impact of the transfer of environmentally friendly technologies from the industrialized world to developing countries—especially when such technologies have a strong economic rationale (World Resources Institute, 1994: 175–78).

One example of the way in which environmental standards within industrialized countries can create spillover benefits for the developing world is the joint venture between the New England Electric Company and Malaysia's Innoprise Corporation, one of the country's major forest products companies which controls 2.5 million acres of Asia's largest remaining continguous rain forest. The goal of the three-year pilot project, monitored by the New York–based Rain Forest Alliance, is to assist Innoprise in adopting low-impact logging techniques (e.g., reducing erosion, preserving surrounding trees, using fewer logging roads, safeguarding forest streams) while improving the forest's yield of commercial-grade timber. New England Electric got involved in the project in an effort to offset its own CO_2 emissions by preserving a portion of Malaysia's carbon-absorbing rain forest. Although the project is purely voluntary in character, New England Electric is betting that it will benefit from future pollution credits in anticipation of legislation regulating carbon dioxide emissions. Moreover, the fact that the Massachusetts Legislature was considering a proposal for a carbon tax of $22 per ton for all CO_2 emissions in the state was a strong motivation for New England Electric to participate in the joint venture. Without the anticipation of new environmental regulations in the United States covering CO_2 emissions, there would be no incentive for New England Electric to become involved in the project (Parrish, 1992).

One can envision a number of institutional reforms that would help amplify the demonstration effect associated with the environmental movement in the First World. For example, Jim MacNeill, Pieter Winsemius, and Taizo Yakushiji (1991) argue that global bargains, involving trades between more developed countries and less developed countries, could provide a foundation for international cooperation on environmental concerns. In a real sense, environmental objectives would serve as the vital medium of exchange. Solar energy technologies, for example, could be traded for programs designed to curb deforestation. Such bargains could be facilitated by the creation of a United Nations Earth Council (the environmental equivalent of the UN Security Council) which would monitor environmental needs geographically and suggest appropriate bargains to petitioning nations. One such global bargain could be the creation of "biosphere trusts" within large rain forests, monitored by a reconfigured UN Trusteeship Council. Developing countries might agree to such trusts if there were sufficient motivations in terms of debt relief or some other financial incentive (Worldwatch Institute, 1992: 171). The Brazilian Tropical Rainforest Fund, established in December 1991 by the Group of Seven industrialized countries with pledges totaling $250 million, hopefully represents a first step toward the concept of biosphere trusts (World Bank, 1992: 177). In addition, with the breakup of the Soviet Union, Eastern Europe offers many exciting opportunities for habitat preservation through the development of international parklands—leveraged through debt-for-nature swaps—along the largely uninhabited lands along the national borders of the former Soviet-dominated states (World Resources Institute, 1992: 59).

The change envisioned by even modest definitions of environmental responsibility have far-reaching implications for the institutions that dominate our domestic and international policy environments. The long and arduous process of securing institutional transformation will demand bold thinking and creative initiative. Most of all, though, it will demand a kind of faith in the ability of humanity to rise to the occasion and to meet the challenges of the twenty-first century, demonstrating courage and insight.

THE COURAGE TO BE HUMAN

To be human is the highest compliment we can be paid. It is not an excuse for weakness, a rationale for a superiority complex, or a justification for pessimism. To be human is to be the recipient of a miraculous gift that must not be taken for granted.

Self-esteem and the ability to believe in oneself are closely related. A person who dislikes himself is a person who finds it difficult to have faith in himself and others. Humans, as a species, have definitely suffered from low self-esteem. We have not expected much of ourselves, and humanity, with the exception of our saints and heroes, has risen to our expectations. Our collective lack of self-respect is demonstrated in many ways—brutality, slavery, power mongering, greed, apathy, obsequiousness, and self-destructive behaviors. Most of all, though, our low opinion of ourselves is reflected in how we keep our home. The carelessness that characterizes our relationship to the environment is but one way of expressing our feelings of worthlessness. What passes for human self-confidence usually has more in common with the fears and insecurities of a playground bully than the confidence that proceeds from an authentic sense of worth and well-being.

Humans are participants in a grand drama that is unfolding before our eyes. We arrived late in the drama . . . so late that we have little understanding of what went on prior to our debut. We, as individuals, have a small part to play in this drama. Some may dismiss it as a bit part, but it is a part nonetheless . . . and the drama is great. Who, after all, can know what may eventually issue from obscurity? Faith beckons us toward risk and understanding. It is the sustenance of life. Faith demands patience, commitment, and vision. Humanity, more than ever, requires all the resources of faith to perform its role in LIFE's drama over the next thirty to forty years. Hopefully, our performance will be remembered admiringly by all who look back. To be human demands nothing less than to live with courage and passion.

NOTES

1. As explained in Chapter 4, I place the term "life" in all capitals to represent the totality of biospheric life as distinct from particular life forms (e.g., chimpanzees, dolphins, humans, spruce trees).

2. This, of course, does not preclude the belief in the existence of an "ultimate end" like God. In fact, the belief in God can be a powerful resource for critiquing the varieties of human purpose.

3. Such integrated conservation-development projects, although promising, have their own set of difficulties. See Brandon and Wells (1992).

4. For a helpful discussion on environmental governance, see Hilary F. French, "Strengthening Global Environmental Governance," Chapter 10 of the 1992 *State of the World* report of the Worldwatch Institute.

5. It should be noted, however, that the protocol lacked an effective incentive structure for accelerating the phaseout of CFCs (see Munasinghe and King, 1992), even though it is a model for environmental negotiations in other respects.

References

Ahmadjian, Vernon, and Surindar Paracer. (1986) *Symbiosis: An Introduction to Biological Associations.* Hanover, NH: University Press of New England.

Anderson, Terry L., and Donald R. Leal. (1991). *Free Market Environmentalism.* Boulder, CO: Westview Press.

Annis, Sheldon. (1992). "Evolving Connectedness among Environmental Groups and Grassroots Organizations in Protected Areas of Central America." *World Development* 20, no. 4, pp. 587–95.

Arrow, Kenneth. (1973). "Some Ordinalist-Utilitarian Notes on Rawls's Theory of Justice." *The Journal of Philosophy* 70, pp. 245–63.

Attfield, Robin. (1983). *The Ethics of Environmental Concern.* New York: Columbia University Press.

Austin, Richard C. (1985). "Beauty: A Foundation for Environmental Ethics." *Environmental Ethics* 7, pp. 197–208.

Balogh, Thomas. (1982). *The Irrelevance of Conventional Economics.* New York: Liveright Publishing Co.

Barash, David P. (1977). *Sociobiology and Behavior.* New York: Elsevier.

Barro, Robert J. (1994). "Federal Protection—Only Cute Critters Need Apply." *Wall Street Journal*, August 4, p. A 14.

Barry, Brian. (1973). *The Liberal Theory of Justice.* Oxford: Oxford University Press.

Bartlett, Ellen. (1991). "Madagascar—Once-idyllic Island in an Ecological Crisis Stripped, Burned; It's Becoming the Ecological Equivalent of the South Bronx." *Boston Globe*, August 12, p. 33.

Belenky, Mary Field, Blythe McVicker Clinchy, Nancy Rule Goldberger, and Jill Mattuck Tarule. (1986). *Women's Ways of Knowing: The Development of Self, Voice, and Mind.* New York: Basic Books.

Bentham, Jeremy. (1789). "An Introduction to the Principles of Morals and Legislation." In *The Collected Works of Jeremy Bentham*, edited by J. H. Burnes and H.L.A. Hart. London: Athlone Press, 1970.

Beyleveld, Deryck. (1991). *The Dialectical Necessity of Morality: An Analysis and Defense of Alan Gewirth's Argument to the Principle of Generic Consistency*. Chicago: University of Chicago Press.

Birch, Charles, and John B. Cobb, Jr. (1981). *The Liberation of Life*. Cambridge: Cambridge University Press.

Booth, Annie L., and Harvey L. Jacobs. (1990). "Ties That Bind: Native American Beliefs as a Foundation for Environmental Consciousness." *Environmental Ethics* 12, no. 1, pp. 27–43.

Brandon, Katrina Eadie, and Michael Wells. (1992). "Planning for People and Parks: Design Dilemmas." *World Development* 20, no. 4, pp. 557–70.

Bratton, Susan Power. (1988). "The Original Desert Solitare: Early Christian Monasticism and Wilderness." *Environmental Ethics* 10, pp. 31–53.

Broad, Robin. (1994). "The Poor and the Environment: Friends or Foes?" *World Development* 22, no. 6, pp. 811–22.

Broad, Robin, and John Cavanagh. (1993). *Plundering Paradise: The Struggle for the Environment in the Philippines*. Berkeley: University of California Press.

Brooke, James. (1993). "Oil and Tourism Don't Mix, Inciting Amazon Battle." *New York Times*, September 26, p. 3.

Brown, Barbara E., and John C. Ogden. (1993). "Coral Bleaching." *Scientific American*, January, pp. 64–70.

Brown, Lester R., Christopher Flavin, and Sandra Postel. (1991). *Saving the Planet: How to Shape an Environmentally Sustainable Global Economy*. New York: W. W. Norton.

Cairncross, Frances. (1992). *Costing the Earth*. Boston: Harvard Business School Press.

Calhoun, Suzanne, and Robert L. Thompson. (1988). "Long Term Retention of Self-recognition by Chimpanzees." *American Journal of Primatology* 15, pp. 361–65.

Callicott, J. Baird. (1980). "The Search for an Environmental Ethic." In *Matters of Life and Death*, edited by Tom Regan, 2d ed. New York: Random House.

———. (1982). "Hume's Is/Ought Dichotomy and the Relation of Ecology to Leopold's Land Ethic." *Environmental Ethics* 4, pp. 163–74.

——— (ed.). (1987). *Companion to A Sand County Almanac: Interpretive and Critical Essays*. Madison: University of Wisconsin Press.

———. (1989). *In Defense of the Land Ethic: Essays in Environmental Philosophy*. Albany: State University of New York.

Callicott, J. Baird, and Roger T. Ames (eds.). (1989). *Nature in Asian Traditions of Thought: Essays in Environmental Philosophy*. Albany: State University of New York.

Capra, Fritjof. (1975). *The Tao of Physics*. Boulder, CO: Shambala Publications.

Coase, Ronald. (1960). "The Problem of Social Cost." *Journal of Law and Economics* 3, pp. 1–45.

Coate, Roger A. (1982). *Global Issue Regimes*. New York: Praeger.

Cochrane, Susan H. (1979). *Fertility and Education: What Do We Really Know?* Baltimore: Johns Hopkins University Press.

Cody, Edward. (1991). "In Mexico City, There's Fear in the Air; Early Data Indicate Capital's Chronic Pollution May Be Worse Than Ever This Winter." *Washington Post*, November 24, p. A27.

Collard, David. (1978). *Altruism and Economy: A Study of Non-Selfish Economics*. New York: Oxford University Press.

Cooter, Robert, and Peter Rappoport. (1984). "Were the Ordinalists Wrong about Welfare Economics?" *Journal of Economic Literature* 22, pp. 507–30.

Costanza, Robert (ed.). (1991). *Ecological Economics: The Science and Management of Sustainability*. New York: Columbia University Press.

Crook, John H. (1980). *The Evolution of Human Consciousness*. Oxford: Clarendon Press.

Daly, Herman E. (1991). *Steady-State Economics*. 2d ed. Washington, DC: Island Press.

Daly, Herman E., and John B. Cobb, Jr. (1989). *For the Common Good*. Boston: Beacon Press.

d'Aquili, Eugene G. (1978). "The Neurobiological Bases of Myth and Concepts of Deity." *Zygon* 13, pp. 257–75.

Dawkins, Richard. (1976). *The Selfish Gene*. New York: Oxford University Press.

———. (1982). *The Extended Phenotype*. Oxford: W. H. Freedman and Co.

———. (1987). *The Blind Watchmaker*. New York: W. W. Norton.

Deane, Phyllis. (1978). *The Evolution of Economic Ideas*. Cambridge: Cambridge University Press.

Diamond, Jared. (1990). "Bach, God, and the Jungle." *Natural History* 99, no. 11, pp. 22–27.

Donagan, Alan. (1977). *The Theory of Morality*. Chicago: University of Chicago Press.

Donald, Merlin. (1991). *Origins of the Modern Mind*. Cambridge, MA: Harvard University Press.

Drogin, Robert. (1990). "Manila—A City out of Control." *Los Angeles Times*, May 29, p. 6(H).

Ehrenfeld, David. (1991). "The Management of Diversity: A Conservation Paradox." In *Ecology, Economics, Ethics: The Broken Circle*, edited by F. Herbert Bormann and Stephen R. Kellert. New Haven, CT: Yale University Press.

Eibl-Eibesfeldt, Irenäus. (1989). *Human Ethology.* New York: Aldine de Gruyter.

Eliade, Mircea. (1954). *The Myth of the Eternal Return.* New York: Pantheon Books.

———. (1959). *The Sacred and the Profane.* New York: Harcourt, Brace and Co.

Elkins, J. W., T. M. Thompson, T. H. Swanson, et al. (1993). "Decrease in the Growth Rates of Atmospheric Chlorofluorocarbons 11 and 12." *Nature* 364 (August 26), pp. 780–83.

Elliot, Robert. (1989). "Environmental Degradation, Vandalism and the Aesthetic Object Argument." *Australian Journal of Philosophy* 67, pp. 191–204.

El Serafy, Salah. (1988). "The Proper Calculation of Income from Depletable Natural Resources." In *Environmental and Resource Accounting and Their Relevance to the Measurement of Sustainable Income,* edited by Ernst Lutz and Salah El Serafy. Washington, DC: World Bank.

———. (1991). "The Environment as Capital." In *Ecological Economics: The Science and Management of Sustainability,* edited by Robert Costanza. New York: Columbia University Press.

Elton, Charles. (1930). *Animal Ecology and Evolution.* Oxford: Oxford University Press.

Etzioni, Amitai. (1988). *The Moral Dimension: Toward a New Economics.* New York: The Free Press.

Farah, Douglas. (1992). "Ecuador Cedes Amazon Land to Indians." *Washington Post,* May 15, p. A30.

Fisher, Julie. (1993). *The Road from Rio: Sustainable Development and the Nongovernmental Movement in the Third World.* Westport, CT: Praeger.

"Food for Thought." (1992). *The Economist,* August 29, pp. 64, 66.

Frankena, William. (1949). "The Naturalistic Fallacy." *Mind* 48, pp. 464–77.

Freed, Kenneth. (1991). "Salvador's Ecological Nightmare." *Los Angeles Times,* June 15, p. 1(A)

Gabor, Dennis. (1972). *The Mature Society.* London: Secker and Warburg.

Gallagher, Winifred. (1993). "Sacred Places." *Psychology Today* 26, no. 1 (January/February), pp. 62–70.

Gallup, G. G., Jr. (1970). "Chimpanzees: Self-recognition." *Science* 167, pp. 86–87.

Gardner, R. A., and B. T. Gardner. (1969). "Teaching Sign Language to a Chimpanzee." *Science* 165, pp. 664–72.

Gewirth, Alan. (1978). *Reason and Morality.* Chicago: University of Chicago Press.

———. (1979). "Starvation and Human Rights." In *Ethics and Problems of the 21st Century,* edited by K. Goodpaster and K. Sayre. Notre Dame, IN: University of Notre Dame Press.

———. (1982). *Human Rights: Essays on Justification and Applications*. Chicago: University of Chicago Press.

Giddens, Anthony. (1979). *Central Problems in Social Theory*. Berkeley: University of California Press.

———. (1984). *The Constitution of Society*. Berkeley: University of California Press.

Gilligan, Carol. (1982). *In a Different Voice: Psychological Theory and Women's Development*. Cambridge, MA: Harvard University Press.

Gunn, Alastair. (1984). "Preserving Rare Species." In *Earthbound: New Introductory Essays in Environmental Ethics*, edited by Tom Regan. New York: Random House.

Hamilton, W. D. (1964) "The Genetical Theory of Social Behavior." *Journal of Theoretical Biology* 7, pp. 1–52.

Hardin, Garrett. (1968). "The Tragedy of the Commons." *Science* 162, pp. 1243–48.

Hardy, John, and Hermann Gucinski. (1989). "Stratospheric Ozone Depletion: Implications for Marine Ecosystems." *Oceanography* (November), pp. 18–19.

Harmon, David. (1987). "Cultural Diversity, Human Subsistence, and the National Park Ideal." *Environmental Ethics* 9, pp. 147–58.

Heuting, Roefie. (1991). "Correcting National Income for Environmental Losses: A Practical Solution for a Theoretical Dilemma." In *Ecological Economics: The Science and Management of Sustainability*, edited by Robert Costanza. New York: Columbia University Press.

Hicks, John. (1941). "The Rehabilitation of Consumers' Surplus." *Review of Economic Studies* 8, pp. 108–16

Hiltzik, Michael A. (1989). "An Urge to Burn; Madagascar: Rare Island up in Smoke." *Los Angeles Times*, May 13, p. 1(1).

Hirsh, Fred. (1976). *Social Limits to Growth*. Cambridge, MA: Harvard University Press.

Hoffman, W. Michael. (1991). "Business and Environmental Ethics." *Business Ethics Quarterly* 1, no. 2, pp. 169–84.

Hotelling, Harold. (1931). "The Economics of Exhaustible Resources." *Journal of Political Economy* 39, pp. 137–75.

Hume, David. (1948a). *Hume's Moral and Political Philosophy*, edited by Henry D. Aiken. New York: Hafner Publishing Co.

———. (1948b). *A Treatise on Human Nature*. New York: AMS Press.

Isaac, Glynn Ll. (1983). "Aspects of Human Evolution." In *Evolution from Molecules to Man*, edited by D. S. Bendall. Cambridge: Cambridge University Press.

Jevons, William Stanley. (1911). *Theory of Political Economy*. 4th ed. London: Macmillan.

Johnson, Lawrence E. (1991). *A Morally Deep World: An Essay on Moral Sig-*

nificance and Environmental Ethics. Cambridge: Cambridge University
    Press.
Kant, Immanuel. (1785). *Groundwork of the Metaphysics of Morals*. Trans. by
    H. J. Paton. New York: Harper and Row, 1964.
————. (1797). *The Metaphysical Principles of Virtue*. Trans. by James Elling-
    ton. Indianapolis, IN: Bobbs-Merrill, 1964.
Katouzian, Homa. (1980). *Ideology and Method in Economics*. New York: New
    York University Press.
Keynes, John Maynard. (1923). "A Tract on Monetary Reform." In *The Col-
    lected Writings of John Maynard Keynes*, vol. IV. London: Macmillan,
    1971.
Korten, David. (1990). *Getting to the 21st Century: Voluntary Action and the
    Global Agenda*. Hartford, CT: Kumarian Press.
Larson, Bruce A. (1994). "Changing the Economics of Environmental Degra-
    dation in Madagascar: Lessons from the National Environmental Action
    Plan Process." *World Development* 22, no. 5, pp. 671–89.
Laughlin, Charles D., and Eugene G. d'Aquili. (1974). *Biogenetic Structuralism*.
    New York: Columbia University Press.
Lea, Stephen E. G., Roger M. Tarpy, and Paul Webley. (1987). *The Individual
    in the Economy: A Textbook of Economic Psychology*. Cambridge: Cam-
    bridge University Press.
Leakey, Richard E., and Roger Lewin. (1977). *Origins*. New York: E. P. Dutton.
————. (1992). *Origins Reconsidered*. New York: Doubleday.
Lele, Sharachchandra M. (1991). "Sustainable Development: A Critical Re-
    view." *World Development* 19, pp. 607–21.
Lemonick, Michael D. (1993a). "The Hunt, the Furor." *Time*, August 2, pp.
    42–45.
————. (1993b). "Secrets of the Maya." *Time*, August 9, pp. 44–50.
Leopold, Aldo. (1933). "The Conservation Ethic." *Journal of Forestry* 31, pp.
    634–43.
————. (1949). *A Sand County Almanac*. New York: Oxford University Press.
Levi-Strauss, Claude. (1963). *Totemism*. Boston: Beacon Press.
————. (1966). *The Savage Mind*. Chicago: University of Chicago Press.
Lewis, David K. (1969). *Convention: A Philosophical Study*. Cambridge, MA:
    Harvard University Press.
Linden, Eugene. (1989). "The Death of Birth." *Time*, January 2, p. 33.
Liotta, Lance A. (1992). "Cancer Cell Invasion and Metastasis." *Scientific
    American*, February, pp. 54–63.
Liotta, Lance A., Patricia S. Steeg, and William G. Stetler-Stevenson. (1991).
    "Cancer Metastatis and Angiogenesis: An Imbalance of Positive and
    Negative Regulation." *Cell* 64, pp. 327–36.
Little, I.D.M. (1957). *A Critique of Welfare Economics*. 2d ed. Oxford: Clar-
    endon Press.
————. (1979). "Welfare Criteria, Distribution, and Cost-Benefit Analysis." In

Economics and Human Welfare, edited by Michael Boskin. New York: Academic Press.

Lorenz, Konrad Z. (1981). *The Foundations of Ethology*. New York: Springer-Verlag.

Lovelock, James. (1979). *Gaia: A New Look at Life on Earth*. Oxford: Oxford University Press.

———. (1990). *The Ages of Gaia: A Biography of Our Living Earth*. New York: Bantam Books.

Lovelock, James, and Sidney Epton. (1975). "The Quest for Gaia." *New Scientist* 6 (February), pp. 304ff.

Lumsden, Charles J., and Edward O. Wilson. (1981). *Genes, Mind and Culture*. Cambridge, MA: Harvard University Press.

———. (1983). *Promethean Fire: Reflections on the Origin of the Mind*. Cambridge, MA: Harvard University Press.

MacArthur, Robert, and Edward O. Wilson (1967). *The Theory of Island Biogeography*. Princeton, NJ: Princeton University Press.

McNeely, Jeffrey A. (1988). *Economics and Biological Diversity*. Gland, Switzerland: International Union for Conservation of Nature and Natural Resources.

McNeely, Jeffrey A., et al. (1990). *Conserving the World's Biological Diversity*. Gland, Switzerland, and Washington, DC: International Union for Conservation of Nature and Natural Resources, World Resources Institute, Conservation International, World Wildlife Fund—U.S., and the World Bank.

MacNeill, Jim, Pieter Winsemius, and Taizo Yakushiji. (1991). *Beyond Interdependence: The Meshing of the World's Economy and the Earth's Ecology*. New York: Oxford University Press.

Magistad, Mary Kay. (1991). "Bangkok's Progress Marked by Health Hazards; Poor Air Quality, Noise Pollution and Traffic Congestion Cause Stress, Diseases." *Washington Post*, May 7, p. Z13.

Mann, Judy. (1992). "Family Planning in the Family of Nations." *Washington Post*, December 2, p. B23.

Maritain, Jacques. (1944). *The Rights of Man and Natural Law*. London: Geoffrey Bles.

———. (1951). *Man and the State*. Chicago: University of Chicago Press.

Martin, Calvin. (1978). *Keepers of the Game: Indian-American Relationships and the Fur Trade*. Berkeley: University of California Press.

Martin, Paul S. (1973). "The Discovery of America." *Science* 179, pp. 969–74.

Marx, Jean. (1993). "Cell Death Studies Yield Cancer Clues." *Science* 259, pp. 760–61.

Masland, Tom. (1989). "Enchanting Forest; South America Isn't the Only Place Where Nature Is Losing Ground." *Chicago Tribune*, May 31, p. 1-C.

Maslow, Abraham H. (1948). "'Higher' and 'Lower' Needs." *Journal of Psychology* 25, pp. 433–36.

Mayr, Ernst. (1970). *Population, Species and Evolution.* Cambridge, MA: Harvard University Press.

―――. (1976). *Evolution and the Diversity of Life.* Cambridge, MA: Harvard University Press.

―――. (1988). *Toward a New Philosophy of Biology.* Cambridge, MA: Harvard University Press.

Mill, John Stuart. (1861). *Utilitarianism.* Edited by Oskar Piest. Indianapolis, IN: Bobbs-Merrill, 1957.

Miller, Peter. (1982). "Value as Richness: Toward a Value Theory for an Expanded Naturalism in Environmental Ethics." *Environmental Ethics* 4, pp. 101–14.

Mishan, Ezra J. (1979). "Does Perfect Competition in Mining Produce an Optimal Rate of Exploitation?" In *Theory for Economic Efficiency,* edited by Harry I. Greenfield. Cambridge, MA: MIT Press.

Moore, George E. (1903). *Principia Ethica.* Cambridge: Cambridge University Press.

Morgan, Elizabeth A., Grant D. Power, and Van B. Weigel. (1993). "Thinking Strategically About Development: A Typology of Action Programs for Global Change." *World Development* 21, no. 12, pp. 1913–30.

Munasinghe, Mohan. (1993). "Environmental Issues and Economic Decisions in Developing Countries." *World Development* 21, no. 11, pp. 1729–48.

Munasinghe, Mohan, and Kenneth King. (1992). "Accelerating Ozone Layer Protection in Developing Countries." *World Development* 20, no. 4, pp. 609–18.

Mussen, Paul H., and Nancy Eisenberg-Berg. (1977). *Roots of Caring, Sharing and Helping.* San Francisco: Freeman.

Myers, Norman. (1991). "Biological Diversity and Global Security." In *Ecology, Economics, Ethics: The Broken Circle,* edited by F. Herbert Bormann and Stephen R. Kellert. New Haven, CT: Yale University Press.

Myrdal, Gunnar. (1953). *The Political Element in the Development of Economic Theory.* London: Routledge and Kegan Paul.

Naess, Arne. (1979). "Self-realization in Mixed Communities of Humans, Bears, Sheep and Wolves." *Inquiry* 22, pp. 231–41.

Nash, Roderick F. (1989). *The Rights of Nature: A History of Environmental Ethics.* Madison: University of Wisconsin Press.

Neill, Robin F. (1978). "The Ethical Foundations of Economics." *Philosophy in Context* 7, pp. 86–95.

Niebuhr, Reinhold. (1932). *Moral Man and Immoral Society.* New York: Charles Scribner's Sons.

Nozick, Robert. (1974). *Anarchy, State and Utopia.* New York: Basic Books.

Oelschlaeger, Max. (1994). *Caring for Creation: An Ecumenical Approach to the Environmental Crisis.* New Haven, CT: Yale University Press.

O'Neill, Onora. (1986). *Faces of Hunger: An Essay on Poverty, Justice and Development.* London: George Allen and Unwin.

Organization for Economic Cooperation and Development. (1991). *Environmental Labelling in OECD Countries.* Paris: OECD.

Parrish, Michael. (1992). "L.A. Firm Helps Utility with Innovative Plan." *Los Angeles Times*, August 4, p. D1.

Patterson, Francine. (1978). "Conversations with a Gorilla." *National Geographic* 154, pp. 438–65.

Peskin, Henry M. (1991). "Alternative Environmental and Resource Accounting Approaches." In *Ecological Economics: The Science and Management of Sustainability*, edited by Robert Costanza. New York: Columbia University Press.

"The Poisoned Giant Wakes Up." (1989). *The Economist*, November 4, pp. 23–26.

Premack, Ann James, and David Premack. (1972). "Teaching Language to an Ape." *Scientific American* 277, no. 4, pp. 92–99.

Protzman, Ferdinand. (1989). "East Berliners Explore Land Long Forbidden." *New York Times*, November 10, pp. A1, A14.

"The Questions Rio Forgets." (1992). *The Economist*, May 30, pp. 11–12.

Raff, Rudolf A., and Thomas C. Kaufman. (1983). *Embryos, Genes, and Evolution.* New York: Macmillan.

Rawls, John. (1971). *A Theory of Justice.* Cambridge, MA: Harvard University Press.

Regan, Tom. (1983). *The Case for Animal Rights.* Berkeley: University of California Press.

Register, Richard. (1992). "Ecological Community Design." In *Sustainable Cities: Concepts and Strategies for Eco-City Development*, edited by Bob Walter, Lois Arkin, and Richard Crenshaw. Los Angeles: Eco-Home Media.

Remnick, David. (1991). "Stalin's Lethal Legacy of Filth, 5-Year Plans; City Lies in Ecological Ruins." *Washington Post*, May 21, p. A1.

Robbins, Lionel. (1932). *An Essay on the Nature and Significance of Economic Science.* London: Macmillan.

Rolston, Holmes, III. (1986). *Philosophy Gone Wild: Essays in Environmental Ethics.* Buffalo, NY: Prometheus Books.

———. (1988). *Environmental Ethics: Duties to and Values in the Natural World.* Philadelphia: Temple University Press.

———. (1991). "Environmental Ethics: Values and Duties in the Natural World." In *Ecology, Economics, Ethics: The Broken Circle*, edited by F. Herbert Bormann and Stephen R. Kellert. New Haven, CT: Yale University Press.

Ross, Nancy Wilson. (1980). *Buddhism: A Way of Life and Thought*. New York: Random House.

Rothenburg, Stephen J., Lourdes Schnaas-Arrieta, Irving A. Perez-Guerrero, et al. (1989). "Evaluación del riesgo potential de la exposición perinatal al plomo en el Valle de Mexico." *Perinatologia y Reproducción Humana* 3, no. 1, pp. 49, 56.

Ruddick, Sara. (1989). *Maternal Thinking: Toward a Politics of Peace*. New York: Ballantine Books.

Sahlins, Marshall. (1972). *Stone Age Economics*. Chicago: Aldine-Atherton.

Samuelson, Paul A. (1947). *Foundations of Economic Analysis*. Cambridge, MA: Harvard University Press.

Schneider, Keith. (1991). "Ozone Depletion Harming Sea Life; Ultraviolet Rays' Damage in Waters of Antarctica Goes Deeper Than Thought." *New York Times*, November 16, p. 6(L).

Schumpeter, Joseph A. (1954). *History of Economic Analysis*. New York: Oxford University Press.

Scitovsky, Tibor. (1941). "A Note on Welfare Propositions in Economics." *Review of Economic Studies* 9, pp. 77–88.

Sen, Amartya K. (1977). "Rational Fools: A Critique of the Behavioral Foundations of Economic Theory." *Philosophy and Public Affairs* 6, 314–44.

———. (1987). *On Ethics and Economics*. New York: Basil Blackwell, 1987.

Shepherd, Gordon M. (1988). *Neurobiology*. 2d ed. New York: Oxford University Press.

Shue, Henry. (1980). *Basic Rights*. Princeton, NJ: Princeton University Press.

Simon, Julian L. (1977). *The Economics of Population Growth*. Princeton, NJ: Princeton University Press.

Singer, Brent A. (1988). "An Extension of Rawls' Theory of Justice to Environmental Ethics." *Environmental Ethics* 10, pp. 217–32.

Singer, Peter. (1975). *Animal Liberation: A New Ethics for Our Treatment of Animals*. New York: Avon Books.

———. (1979). "Not for Humans Only: The Place of Nonhumans in Environmental Issues." In *Ethics and Problems of the 21st Century*, edited by Kenneth Goodpaster and Kenneth Sayre. Notre Dame, IN: University of Notre Dame Press.

———. (1981). *The Expanding Circle: Ethics and Sociobiology*. New York: Farrar, Straus, and Giroux.

Sjoberg, Lennart. (1989). "Global Change and Human Action: Psychological Perspectives," *International Social Sciences Journal* 121, pp. 413–32.

Smith, Nigel. (1983). "Enchanted Forest." *Natural History* 92 (August), pp. 14, 18–20.

The South Commission. (1990). *The Challenge to the South: The Report of the South Commission*. Oxford: Oxford University Press.

Spitler, Gene. (1985). "Do We Really Need Environmental Ethics?" *Environmental Ethics* 7, pp. 91–92.

Stackhouse, John. (1993). "Soaring Population Portends Hungry Future for Africa." *Chicago Tribune*, May 14, p. 26.

Stammer, Larry B. (1992). "Tourism—Nature's New Ally?" *Los Angeles Times*, April 29, p. 1A.

Standing Bear, Luther. (1933). *Land of the Spotted Eagle*. Lincoln: University of Nebraska Press.

Stenhouse, David. (1974). *The Evolution of Intelligence*. London: George Allen and Unwin.

Sugden, Robert, and Alan Williams. (1978). *The Principles of Practical Cost-Benefit Analysis*. Oxford: Oxford University Press.

Thurow, Lester T. (1984). *Dangerous Currents: The State of Economics*. New York: Vintage Books.

Tinbergen, Niko. (1973). *The Animal in Its World*. Vol. 2. Cambridge, MA: Harvard University Press.

Tobey, James A. (1993). "Toward a Global Effort to Protect the Earth's Biological Diversity." *World Development* 21, no. 12, pp. 1931–45.

Toufexis, Anastasia. (1992). "Endangered Species." *Time*, April 27, pp. 48–50.

Turnbull, Colin M. (1972). *The Mountain People*. New York: Simon and Schuster.

United Nations. (1991). *World Urbanization Prospects, 1990*. New York: United Nations.

Walter, Bob. (1992). "Gardens in the Sky." In *Sustainable Cities: Concepts and Strategies for Eco-City Development*, edited by Bob Walter, Lois Arkin, and Richard Crenshaw. Los Angeles: Eco-Home Media.

Warnock, Mary. (1978). *Ethics since 1900*. 3d ed. Oxford: Oxford University Press.

Weigel, Van B. (1989). *A Unified Theory of Global Development*. New York: Praeger.

Weisman, Alan. (1990). "The Real Indiana Jones and his Pyramids of Doom." *Los Angeles Times Magazine*, October 14, pp. 13–20, 39–42.

Whitehead, Alfred N. (1929). *Process and Reality*. New York: Macmillan.

Wilson, Edward O. (1975). *Sociobiology: The New Synthesis*. Cambridge, MA: Harvard University Press.

———. (1978). *On Human Nature*. Cambridge, MA: Harvard University Press.

———. (1984). *Biophilia*. Cambridge, MA: Harvard University Press.

———. (1991a). "Biodiversity, Prosperity, and Value." In *Ecology, Economics, Ethics: The Broken Circle*, edited by F. Herbert Bormann and Stephen R. Kellert. New Haven, CT: Yale University Press.

———. (1991b). "Rainforest Canopy: The High Frontier." *National Geographic* 180, no. 6 (December), pp. 78–107.

————. (1992). *The Diversity of Life*. Cambridge, MA: Harvard University Press.

————. (1993). "Is Humanity Suicidal?" *New York Times Magazine*, May 30, pp. 24–29.

Worland, Stephen. (1967). *Scholasticism and Welfare Economics*. Notre Dame, IN: University of Notre Dame Press.

World Bank. (1984). *World Development Report, 1984*. New York: Oxford University Press.

————. (1992). *World Development Report, 1992*. New York: Oxford University Press.

World Resources Institute. (1990). *World Resources, 1990–91*. New York: Oxford University Press.

————. (1992). *World Resources, 1992–93*. New York: Oxford University Press.

————. (1994). *World Resources, 1994–95*. New York: Oxford University Press.

Worldwatch Institute. (1990). *The State of the World, 1990*. New York: W. W. Norton.

————. (1991). *The State of the World, 1991*. New York: W. W. Norton.

————. (1992). *The State of the World, 1992*. New York: W. W. Norton.

————. (1993). *The State of the World, 1993*. New York: W. W. Norton.

Yablokov, Alexei, Sviatoslav Zabelin, Mikhail Lemeshev, et al. (1991). "Russia: Gasping for Breath, Choking in Waste, Dying Young." *Washington Post*, August 18, p. C3.

Zaehner, R. C. (1961). *The Dawn and Twilight of Zoroastrianism*. New York: G. P. Putnam's Sons.

Index

About the Author

VAN B. WEIGEL is Associate Professor of Ethics and Economic Development at Eastern College in Pennsylvania. He is the author of *A Unified Theory of Global Development* (Praeger, 1989), a *Choice* outstanding Academic Book.